Solve This

Math Activities for Students and Clubs

Typeset in PageMaker 6.5 by Beverly Ruedi
with the assistance of Andrew Alan Ruedi.

Library of Congress Catalog Card Number 2001089229

ISBN 0-88385-717-0

Printed in the United States of America

Current printing (last digit):
10 9 8 7 6 5 4 3

Solve This

Math Activities for Students and Clubs

James S. Tanton
The Math Circle
Boston

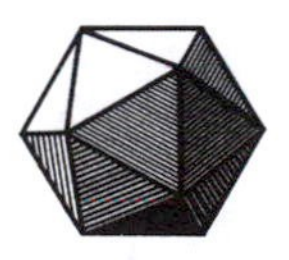

Published and Distributed by
The Mathematical Association of America

CLASSROOM RESOURCE MATERIALS

This series provides supplementary material for students and their teachers—laboratory exercises, projects, historical information, textbooks with unusual approaches for presenting mathematical ideas, career information, and much more.

101 Careers in Mathematics, edited by Andrew Sterrett
Archimedes: What Did He Do Besides Cry Eureka?, Sherman Stein
Calculus Mysteries and Thrillers, R. Grant Woods
Combinatorics: A Problem Oriented Approach, Daniel A. Marcus
A Course in Mathematical Modeling, Douglas Mooney and Randall Swift
Cryptological Mathematics, Robert Edward Lewand
Elementary Mathematical Models, Dan Kalman
Geometry From Africa: Mathematical and Educational Explorations, Paulus Gerdes
Interdisciplinary Lively Application Projects, edited by Chris Arney
Laboratory Experiences in Group Theory, Ellen Maycock Parker
Learn from the Masters, Frank Swetz, John Fauvel, Otto Bekken, Bengt Johansson, and Victor Katz
Mathematical Modeling in the Environment, Charles Hadlock
A Primer of Abstract Mathematics, Robert B. Ash
Proofs Without Words, Roger B. Nelsen
Proofs Without Words II, Roger B. Nelsen
A Radical Approach to Real Analysis, David M. Bressoud
She Does Math!, edited by Marla Parker
Solve This: Math Activities for Students and Clubs, James S. Tanton

MAA Service Center
P.O. Box 91112
Washington, DC 20090-1112
1-800-331-1MAA FAX: 1-301-206-9789

Introduction

The problem with the subtitle "Math Activities for Students and Clubs" is that potential readers may mistakenly think this book is solely for students in focussed mathematics programs, and for their instructors and teachers. This is far from the case. Although the book sprang from my work running the mathematics club at St. Mary's College of Maryland and at Merrimack College, its spirit (and the spirit of my clubs) is one of opening up the world of mathematics to anyone interested—it is my firm belief that mathematics is accessible *to all*. Showing this accessibility is the goal of the book.

Mathematics is creative, interactive and alive. It certainly can challenge or confound people at any level of training. Mathematics often demands critical thinking, ingenuity and innovation. But these are skills we all already possess. In living our everyday lives, we are often presented with open-ended or ill-defined problems. We are challenged to think clearly and creatively and to shift perspectives; that is, to think like a mathematician. Everyone can do math.

Students in a traditional classroom setting often don't have the opportunity either to witness or to experience for themselves the creative aspect to mathematics (this is often dramatically the case in high school and lower-level college classes; how can these students know if they want to pursue mathematics further?). A crucial aim of the activities in this book is to foster original inquiry, to transform the notion of "solved" from one of completion and closure to one of a new opportunity for continued exploration and creative endeavor.

Sophisticated mathematics can be appealing, accessible, hands-on and, quite simply, fun, even for people hesitant about entering the world of mathematical thought! The popularity of the activities described in this text is testimony to this. Self-described "math phobics" as well as school students, college math majors and faculty have all had fun with these activities, and attained a sense of accomplishment and satisfaction from them.

The following thirty chapters represent my most successful math club activities thus far. I have also used many of these activities in middle school and high school mathematics enrichment programs, incorporated them into my college mathematics courses, used them in young children's workshops, and peformed them as party tricks at social gatherings! Each chapter is a collection of mathematical problems and activities linked by a common theme. They all involve working with objects from our everyday experience (bagels, bicycles, teacups and string, paper and scissors, square grids on a tiled floor, for example). Group participation is strongly encouraged. How to carry out these activities with friends is apparent from their descriptions.

I have attempted to write the text in an easy-to-read and supportive style and have used photographs and hand-drawn diagrams to give it a friendly feel. Most chapters are a blend of known material, classic and new interpretations, and original twists. Apart from being fun, the activities in this book touch on a wide variety of topics and are mathematically deep. Placing a hands-on interpretation to a cerebral idea can be quite a challenge. To me, a good problem possesses an immediate hands-on hook, and it can be generalized or connected to larger mathematical concepts and linked to other problems. All the mathematics is explained in an easy style (usually as a *Note on ...*). Further discourse on some topics (optional reading) requires a knowledge of calculus. All sources and references are given at the end of each chapter.

Each chapter of problems fits comfortably within a 60–90 minute period, even allowing for general discussion and generation of new open questions. The reader or activity leader is of course free to select isolated problems for shorter activity periods, or combine chapters; however, the instructor should not be surprised if student excitement carries enough momentum to extend a single problem discussion into a full 90 minute activity unto itself! Each chapter stands alone, and the chapters need not be read in the order presented.

The mathematical process is one of investigating and "unfolding" sequential layers of depth, finding new perspectives and new applications. To reflect this, I have divided the book, and subsequently each chapter, into three sections:

Part I: Activities, Discussions and Problem Statements

Part II: Hints, Some Solutions and Further Thoughts

Part III: Solutions and Discussions

Instructors reading this text in order to lead a group activity should read all three sections of a selected chapter to prepare for the session. While I don't suggest inhibiting free-form thinking, it is important to have a sense of how and where to direct the group's thoughts. The true aim of an activity may not be fully revealed, for example, until the solutions of a series of preliminary problems are understood. Also, from a practical viewpoint, one needs to be aware of all the supplies needed for an activity (handouts of diagrams, markers, etc.).

Following most problem statements or activities there is a discussion on taking the problem further (*Taking It Further*) or a challenge. This is designed to illustrate how to interpret the problem as a window to new questions and new depths for further thought. These discussions also offer specific guidance to finding new directions of exploration and creative endeavor. All *Taking It Further* challenges are solved in the text, whereas the *Challenges* remain open for personal thought or group discussion. These questions, although at times difficult to answer, can serve as inspiration for student projects. Some *Hard Challenges* are problems still unsolved today.

My most successful math clubs were all well-advertised, were held in a comfortable environment, and included a good selection of problems and activities. Before every session my students and I place bright, easy-to-read signs advertising the club across the campus to attract people from all disciplines. It is hard to work past the stigma that a math club is for math majors only, but we try hard. (See Section 5.2 for an eye-catching way to advertise your event!) We also give our sessions titles designed to invoke curiosity (have a look at some of the titles in the table of contents). We hold our club activities in a location with lots of table room and access to a chalk board, but still with an intimate feel; a small group in a large room feels isolated. We also provide snacks to help break down social and academic barriers, to set people at ease, and just for fun. (Pizza at lunchtime is a good way to attract student participation!) It is of course important to know everybody's name.

All the activities outlined in this text have a significant hands-on component. They are designed so that every participant, even those not comfortable with mathematics, can readily participate. Essentially no preparatory learning is needed to dive right in. Always make sure there are plenty of supplies and hand-outs to share so no-one feels left out. Encourage open conversation and exploration of all questions raised. However, there is the challenge to also keep things moving. I don't hesitate to give big hints and ideas, even giving away the answers. It is imperative that the setting feel non-threatening to all participants and not at all exam-like. No student should ever feel on the spot.

Mathematics is fun. It is challenging, creative and alive. Whether you are a student or an instructor, a "math phobic" or a general reader following your sense of curiosity, I hope the activities described in this book offer you a glimpse of the fascination and delight I experience in thinking about mathematics.

Special thanks to Mike Murrow for the photographs taken at St. Mary's College of Maryland, and to Kevin Salemme for those taken at Merrimack College. Thanks also to Lane Anderson and Chuck Adler for their guidance and assistance in setting up the video feedback activity, to Don Albers for his encouragement and to my wife Lindy for her assistance in editing the manuscript and wonderful support. Most especially thanks to my students for their enthusiasm and willingness to learn. It has been my privilege to work with each of you and my pleasure to know each of you.

Contents

To Lindy and Turner

PART I

ACTIVITIES AND PROBLEM STATEMENTS

1

Distribution Dilemmas

1.1 A Shepherd and his Sheep

Here is a classic puzzle.

An elderly shepherd died and left his entire estate to his three sons. To his first son, whom he favored the most, he bequeathed $\frac{1}{2}$ his flock of sheep, to the second son $\frac{1}{3}$, and to the third son, whom he liked the least, $\frac{1}{9}$ of his flock. (Is there a problem with these proportions?)

Not wishing to contest their father's will, the three sons went to the pasture to begin divvying up the flock. They were alarmed to count a total of 17 sheep! Is there a means for the three sons to carry out their father's wishes successfully?

Taking it Further. Meanwhile, three daughters of a recently deceased shepherdess faced a similar dilemma. Their mother, very wealthy, but also possessing a flawed understanding of fractions, had bequeathed her estate of 495 sheep according to the proportions $\frac{1}{5}$ to her first daughter, $\frac{1}{33}$ to her second, and $\frac{1}{2145}$ to her third! Can her will be successfully honored?

1.2 Iterated Sharing

A group of friends sits in a circle, each with a pile of wrapped candies. (Wrapped candy is used because each piece will be handled by many people before being eaten.) Some people have 20 or more pieces, others none, and the rest some number in between. The distribution is quite arbitrary except for the fact that everyone has been given an *even* number of pieces. A reserve supply is set aside.

The friends now follow these instructions: Give half your candy to the person on your left (and hence receive a supply of candy from the person on your right). Do this simultaneously. Now recount your candy supply. If you now have an odd number of pieces, take an extra piece of candy from the reserve supply. This boosts your pile back up to an even number of pieces and everyone is ready to perform the maneuver again.

What happens to the distribution of candy among these friends if this maneuver is performed over and over again? Will people be forever taking extra pieces from the center, so everyone's amount of candy will grow without bound? Or will the distribution stabilize or equalize in some

124 wrapped candies are distributed with a reserve supply of 100 placed in the center.

sense? Will one person end up with all the candy? Might "clumps" of candy move around the circle with each iteration or some strange oscillatory pattern emerge? Is it possible to predict what the result will be?

Taking it Further. What happens if instead of adding pieces, you *eat* any odd piece of candy to bring your pile back *down* to an even number? What happens if the sharing pattern is varied; say, you all give half your candy to the person on your left and the other half to the person on your right?

2

Weird Shapes

2.1 Plucky Perimeters

These figures share a curious property. What is it?

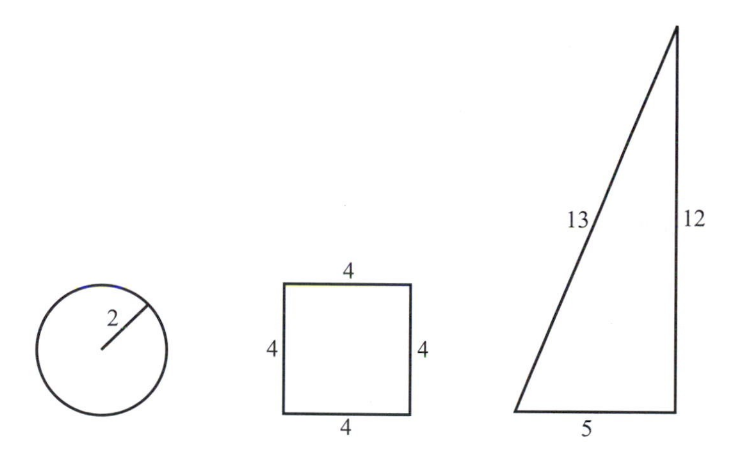

2.2 Weird Wheels

A circular wheel has constant height as it rolls along the ground.

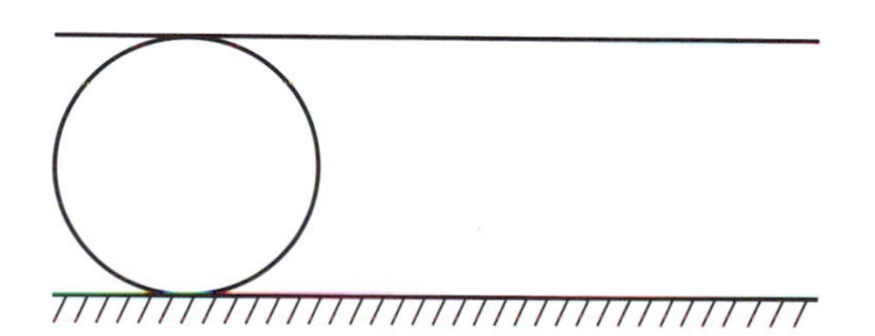

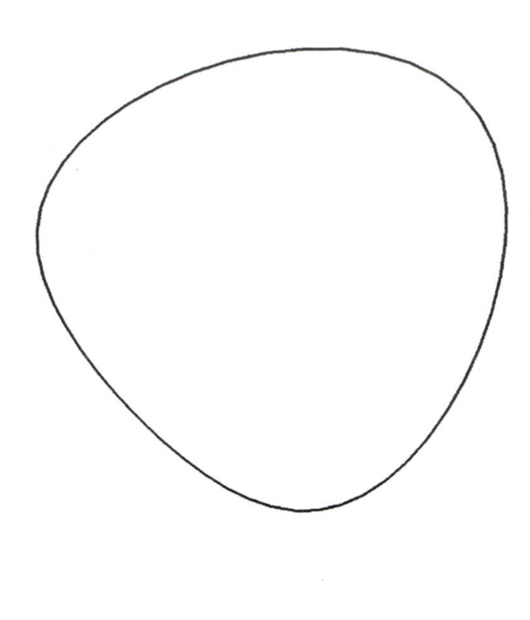

Angela Dellano examines the weird wheel.

Photocopy the irregular shape shown, enlarge the picture, and trace that shape on cardboard to make a wheel. Verify that this wheel also has constant height as it rolls along the ground. How did I make this shape?

2.3 Square Pegs and Not-so-Round Holes

You might want to read the solution to the previous problem before attempting this one.

Square pegs can fit in round holes! For a snug-fitting square peg, all four corners just touch the walls of the hole. A circular hole has the property that this is the case no matter how the square peg is oriented: all four corners always just touch. Is a circle the only shape of a hole that accommodates square pegs in this way?

3

Counting the Odds ... and Evens

3.1 A Coin Trick

Han tosses 12 coins onto a table top. He closes his eyes and instructs John to turn over as many coins as he likes. John can, if he wishes, turn over the same coin every time or any number of times, but there is one proviso: Every time a coin is turned John must say out loud the word "flip."

When finished, John covers one coin with his hand and tells Han it is okay to open his eyes. Han then swiftly, and correctly, announces the state of the coin under John's hand, whether it is heads up or tails up. Han is able to do this correctly every time the game is played, even if a different number of coins is used. What is Han's trick?

Comment. When performing this trick in front of a large group, consider using chips colored black on one side and white on the other rather than coins for better visibility.

3.2 Let's Shake Hands

With an odd number of people in the room, ask everyone to shake hands an odd number of times. No person need shake everyone's hand. In fact, each person could just shake hands with the same small selection of people multiple times. All that is required is that every person be involved in an odd number of handshakes. Noting that it is impolite to refuse a handshake when offered (and that shaking hands with yourself is considered invalid), what curious predicament do folks find themselves in when they attempt this experiment?

Comment. To ensure an odd number of people are involved in this exercise you, as leader, can participate, or not. But keep the motive for your involvment (or lack of involvement) secret!

3.3 Forty-five Cups

Forty-five plastic cups are placed upright on a table top. Turning over six at a time (no more, no less) can you flip all the cups upside down? Try it! (You may have to invert the same cups multiple times in order to accomplish this feat.)

3.4 More Plastic Cups

Twenty-six plastic cups are placed in a row upside down on a table top. Angela turns over every cup. Barry then comes along and turns over every second cup, followed by Cane who turns over every third cup, and so on, all the way down to Zachary who turns over every 26th cup—that is, just the last one!. At the end of this process, which cups are left upright?

John Dockstader attempts to invert 45 cups.

4

Dicing, Slicing, and Avoiding the Bad Bits

4.1 Efficient Tofu Cutting

We can subdivide a cube of tofu into 27 smaller cubes with six planar cuts. Is it possible to complete the same task with fewer than six cuts if we allow stacking the pieces of tofu and slicing through entire stacks with planar cuts?

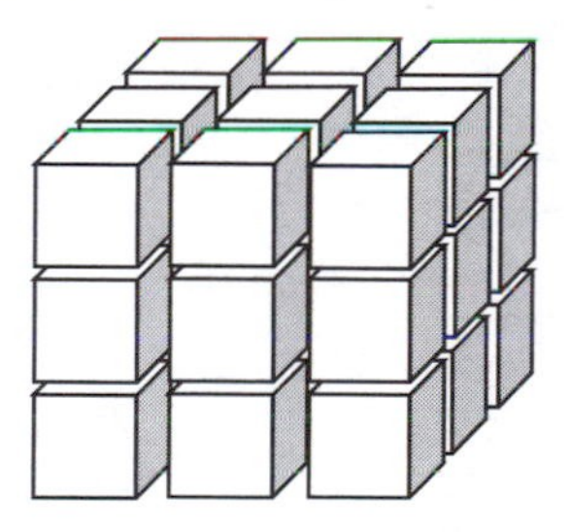

Taking it Further. What is the minimal number of planar cuts needed to dice a $4 \times 4 \times 4$ cube of tofu into 64 smaller cubes? What about a $5 \times 5 \times 5$ cube?

James Taylor attempts to dice a cube of tofu with fewer than six cuts.

4.2 Efficient Paper Slicing

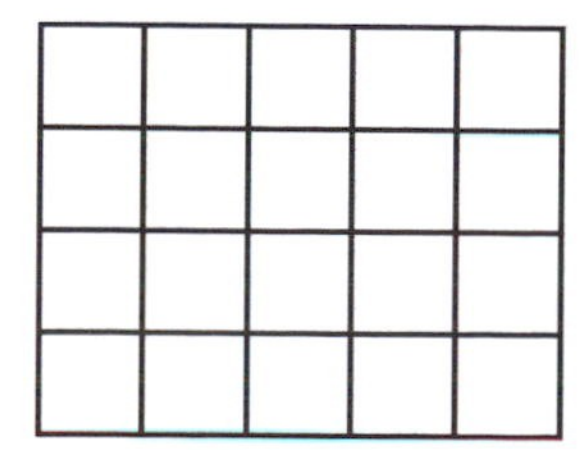

Let's take the tofu cube problem down a dimension: It is possible to slice a $4'' \times 5''$ piece of paper into 20 unit squares with just seven straight line cuts. The same feat can be accomplished with fewer straight line slices if we stack cut pieces of paper during the slicing process (see diagram on p. 10). What is the minimal number of slices required to completely slice a $4'' \times 5''$ piece of paper? What about an arbitrary $n'' \times m''$ piece of paper?

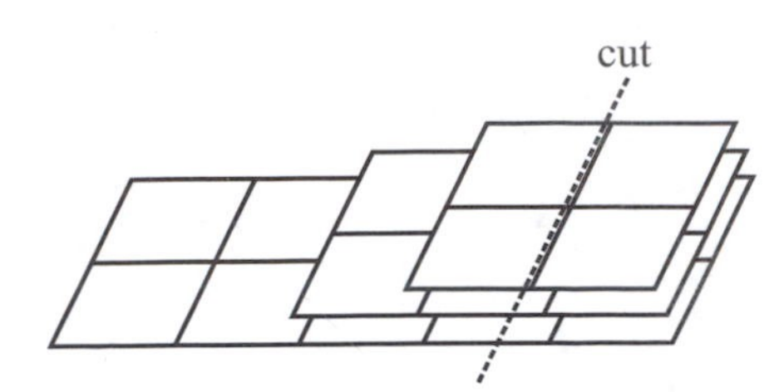

Stacking cut pieces of paper during the slicing process allows you to slice a 4″ × 5″ piece of paper into 20 unit squares with fewer than seven straight line cuts.

4.3 Bad Chocolate (Impossible!)

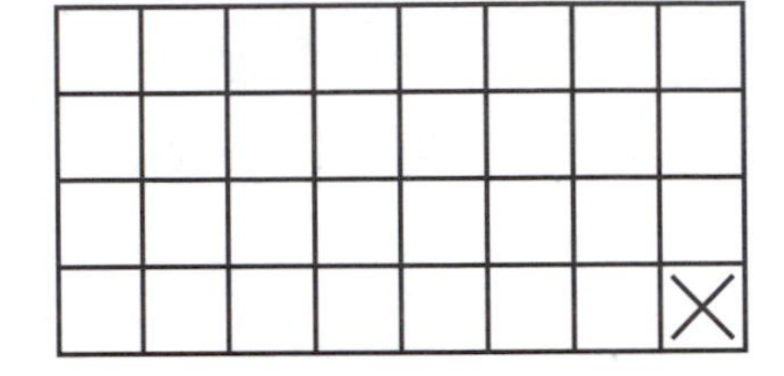

Dan and James are presented with a rectangular 4 × 8 chocolate bar with score marks for breaking it into 32 individual square pieces. They note that the bottom right square of chocolate is spoiled and cannot be eaten.

These gentlemen decide to play the following game: Dan will break the bar along one entire score line, hand the piece containing the bad square to James, and place the remaining piece aside. James will then break the piece handed to him into two sections, again along an entire score line, and hand the portion containing the bad square to Dan, placing the other piece aside; and so on. They will do this until someone is handed a lone square of bad chocolate. That person will then be declared the loser and will keep only the single rotten piece of chocolate, while the other person gets all the rest to enjoy. If you were to play this game, what strategy would you employ?

Taking it Further. A second chocolate bar of different dimensions has a bad square located as shown. Would you want to play the game with this bar?

Kelly Ogden and Angela Dellano of Merrimack College play the bad chocolate game.

5

"Impossible" Paper Tricks

5.1 A Big Hole

Can you cut a hole in an index card big enough to walk through? I can!

5.2 A Mysterious Flap

Paper is manufactured flat and two-dimensional. How then is it possible to construct a piece of paper with a flap extending into the third dimension?

The flap is contiguous with the planar base of paper. No adhesive was used to make this configuration.

5.3 Bizarre Braids

One usually makes a braid with three strands that are joined at one end but free at the other.

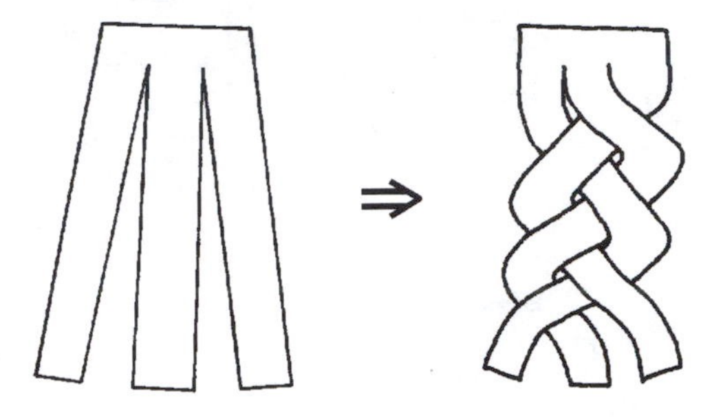

Is it possible to make a braid with no free ends?

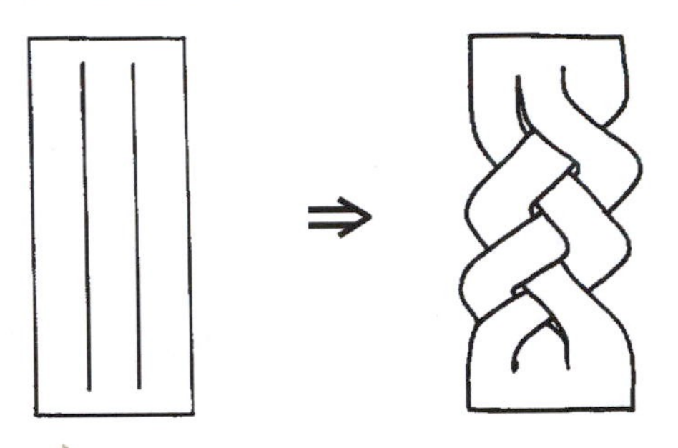

Comment. You can use a slit rectangle of paper, but felt is more flexible and easier to manipulate.

5.4 Linked Unlinked Rings

Two linked rings have the property that if you cut either of them, the configuration falls into two separate pieces.

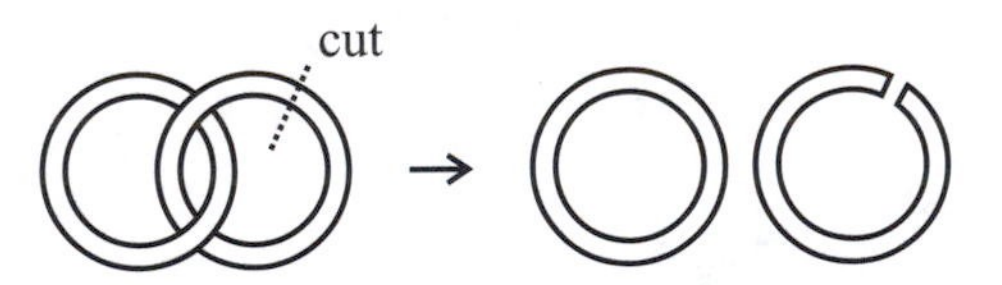

Is it possible to interlink three rings of paper in such a way that if you cut any one of them (either the first, second, or third), just once, the configuration is guaranteed to fall into three separate pieces?

6

Tiling Challenges

6.1 Checkerboard Tiling I

It is impossible to tile a 7×7 grid of squares with 2×1 dominoes in such a way that each square of the grid is covered by one domino and no domino hangs over the edge of the diagram. (Why?) However, if we excise one corner of the grid the surviving configuration of 48 squares is tilable.

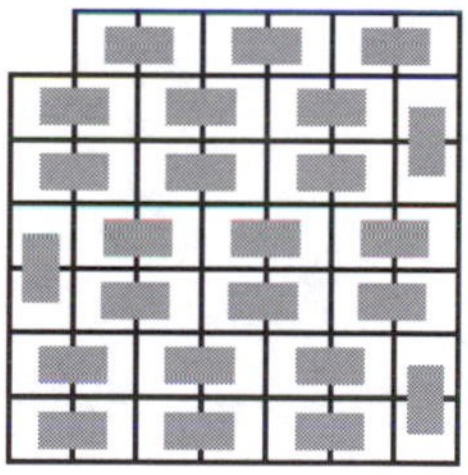

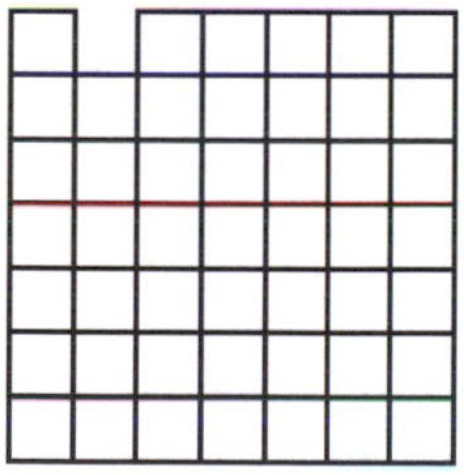

Suppose instead we excise the cell next to the corner. Does this also leave a tilable configuration? (Try tiling it! Draw a grid of squares on paper and use paper clips as dominoes.)

Precisely which cells can be excised from the 7×7 square grid to leave a tilable arrangement of 48 squares?

6.2 Checkerboard Tiling II

This is a classic puzzle from domino tiling theory. If you have worked through section 6.1, its solution should be straightforward.

Two diagonally opposite corners of an 8×8 checkerboard have been excised. Is it possible to tile the remaining configuration of 62 squares with 31 dominoes?

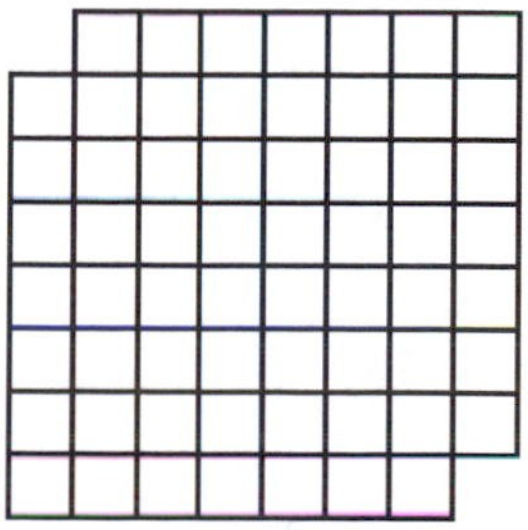

6.3 Checkerboard Tiling III

Is it possible to tile an 8×8 grid of squares with 21 3×1 tiles and one 1×1 tile? If so, how? If not, why not?

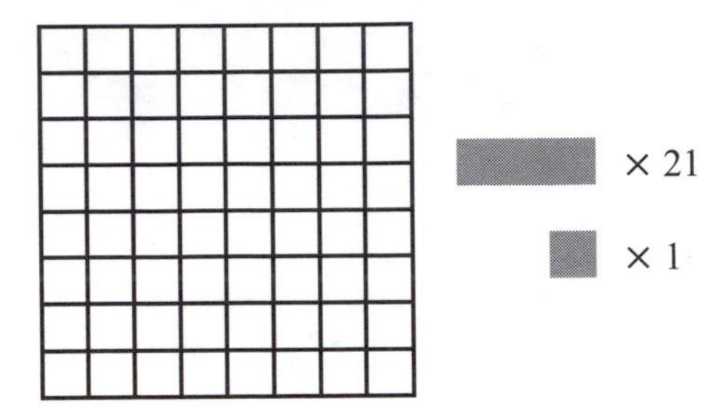

6.4 Checkerboard Tiling IV

Here is a tiling of a 6×6 square array with 18 2×1 dominoes. Notice that this diagram contains a horizontal line that separates the tiles into two disjoint groups (it also contains a vertical line with this property.) Present a tiling of the 6×6 array that avoids such separating lines.

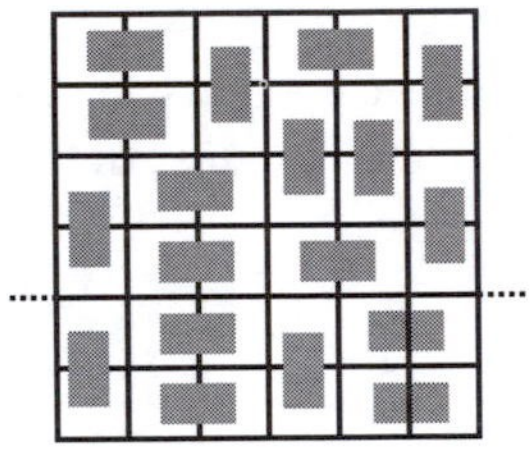

7

Things That Won't Fall Down

7.1 Wildly Wobbly

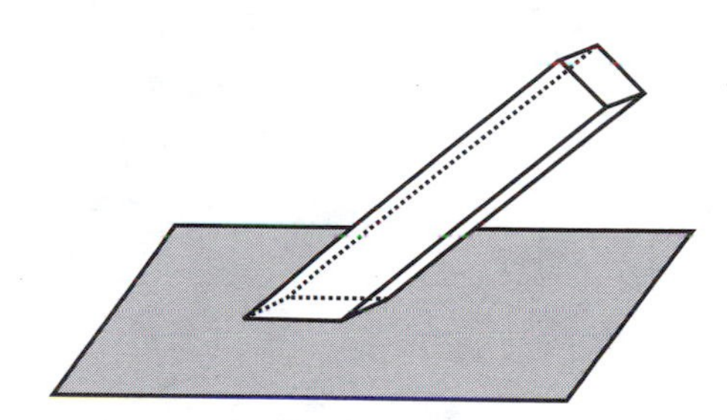

This six-faced polyhedral figure will surely fall over when placed on a table top as shown. Is it possible to design a polyhedral figure that will always topple over, no matter on which face it is placed?

7.2 A Troubling Mobile

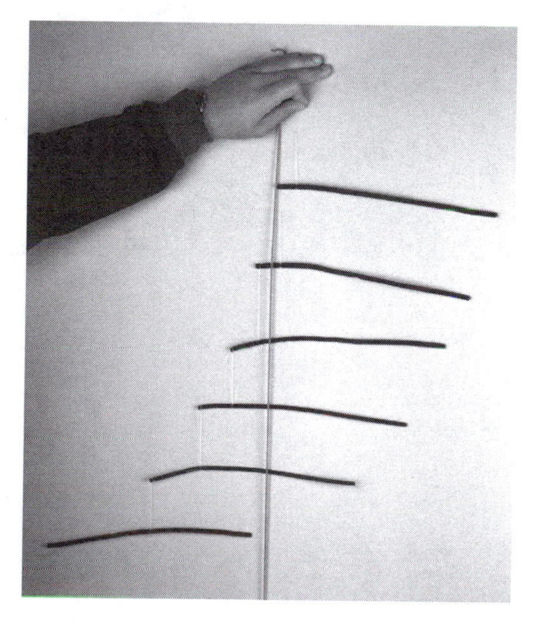

How do you make a perfectly balanced mobile with the property that the lowest wire completely extends beyond the length of the top wire? The mobile shown was made with pipe cleaners and thread (florist wire also works well). Notice that no part of the bottom wire is beneath the top wire.

Notice that the end points of the wires in the photograph form a beautiful curve. Approximately what curve is it?

Taking it Further. Is it possible to make a perfectly balanced mobile with the property that the lowest wire extends outwards by more than *two* lengths?

7.3 A Troubling Tower

It is possible to stack wooden building blocks in a staircase fashion so that the top block completely extends beyond the end point of the bottom block. How?

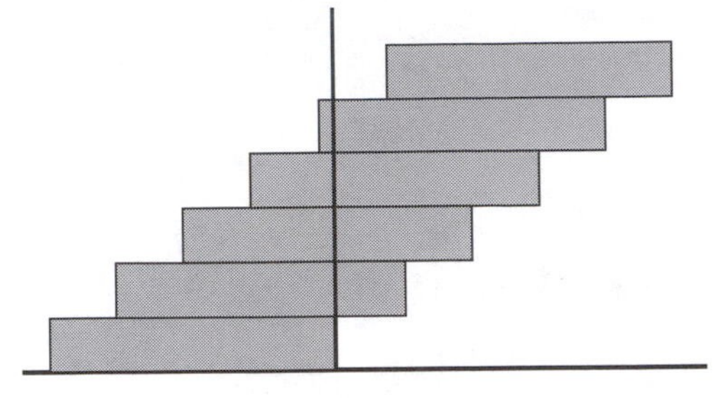

Comment. Yard sticks, cassette tapes and even playing cards also work well for this demonstration.

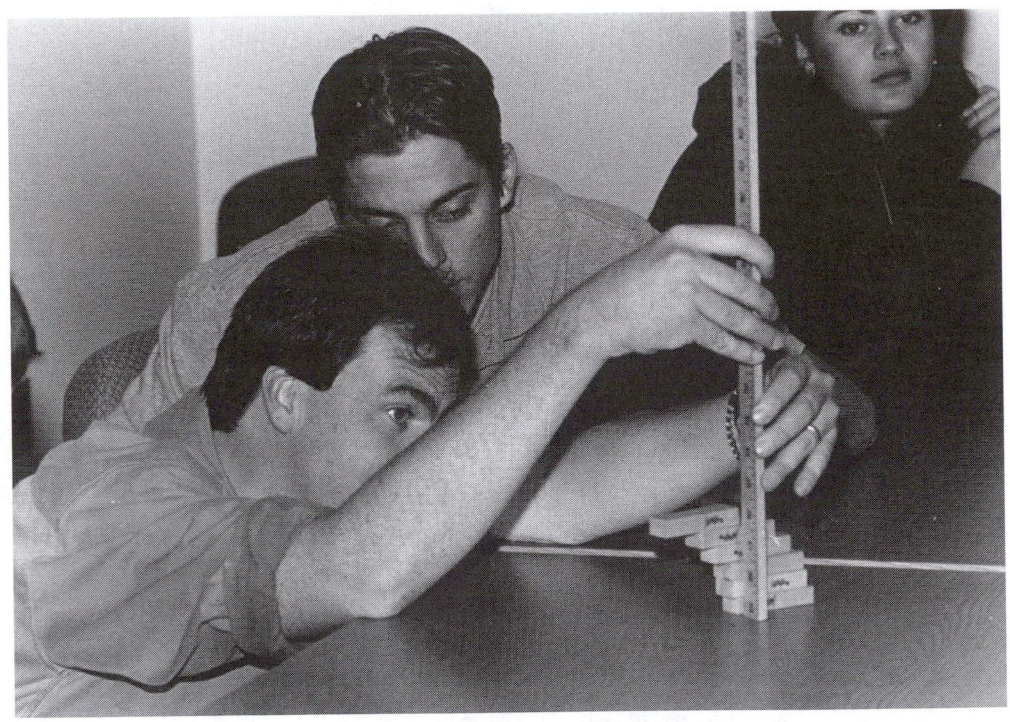

I check Josh Davis' impossible tower.

8

Möbius Madness: Tortuous Twists on a Classic Theme

8.1 Möbius Basics

If a band of paper is cut along its center line it will separate into two pieces. No surprises here!

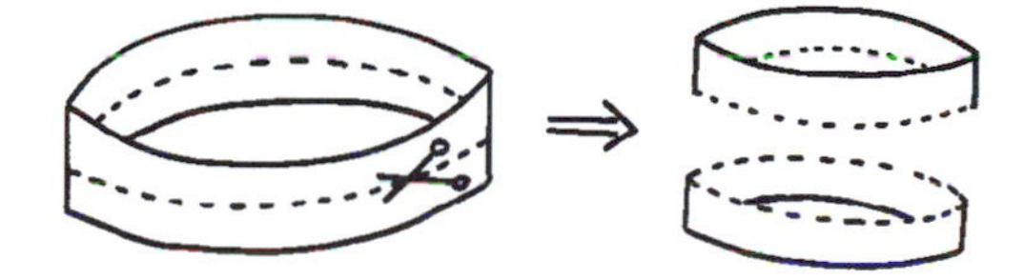

Now take a strip of paper, draw the center line on both sides and form a *Möbius band* by taping the ends together with a half twist. What happens if you cut this figure in half?

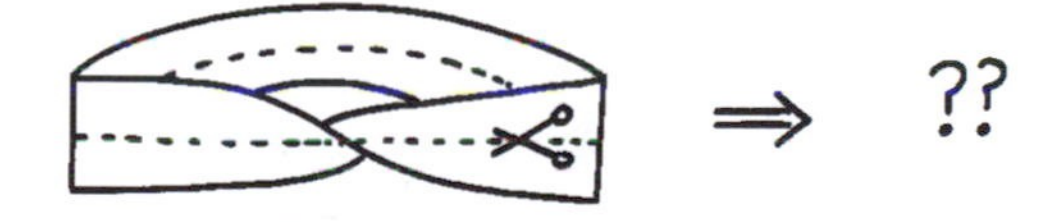

Taking It Further 1. Suppose you cut in half a band with two, three, or more half twists? Can you predict what will result?

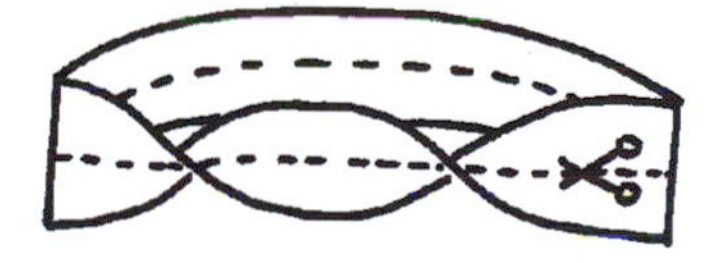

Taking It Further 2. How many pieces result if a Möbius band is cut into thirds? How many pieces result if a band with five half twists is cut into fifths?

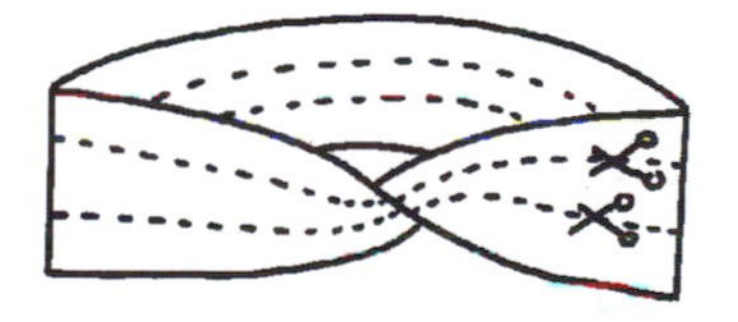

Taking It Further 3. Take a long strip of paper and bring the ends together, with a half twist, to form a Möbius band. But before taping the ends together slide one end of the paper along the strip all the way round back to the other end to form a "double layered" Möbius band. (Equivalently, lay two strips of paper on top of one another, simultaneously give them a half twist, and tape their respective ends together.) What happens if this object is cut along its center line?

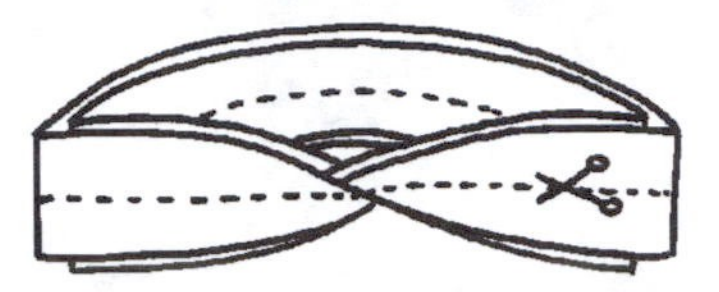

8.2 A Diabolical Möbius Construction

Each of these figures has a center hole. What happens when you cut around the center hole of this figure?

What about in this predicament, with the two half twists in the same direction?

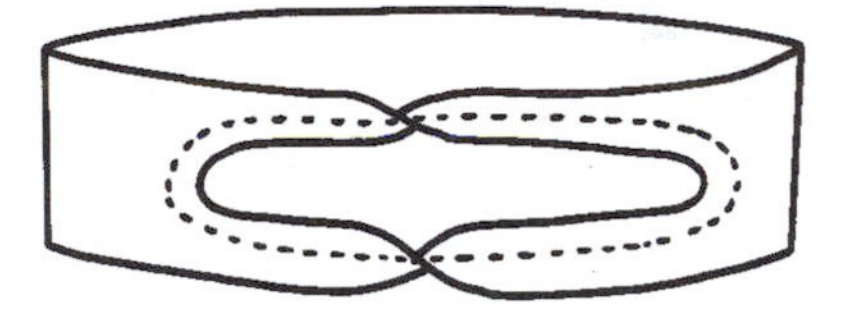

Or this one, with the two half twists in oppositie directions?

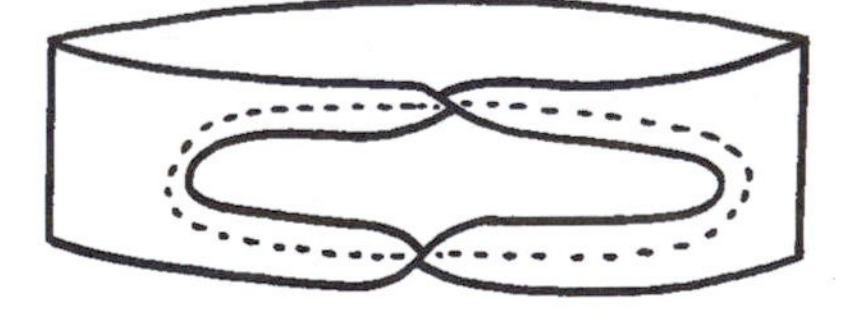

Taking It Further. Experiment with different configurations of twists and multi-twists in the same and opposite directions. Explore what happens and see if you can explain any patterns that occur.

Comment. To make these curious bands, begin with a long wide strip of paper, and cut out half ovals from both ends. Draw guide lines for cutting around each half oval, on both sides of the paper. Then roll up the paper and tape the appropriate ends together with the desired number of half twists.

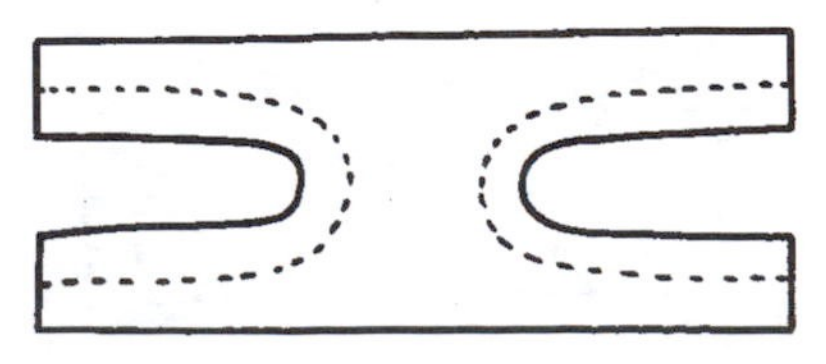

8.3 Another Diabolical Möbius Construction

Cut a piece of paper into an **X** shape to construct two bands of equal length and width attached perpendicularly to one another. What happens when each is cut along its center line?

How does the result change when one band is given a half twist? Two half twists? Seven half twists?

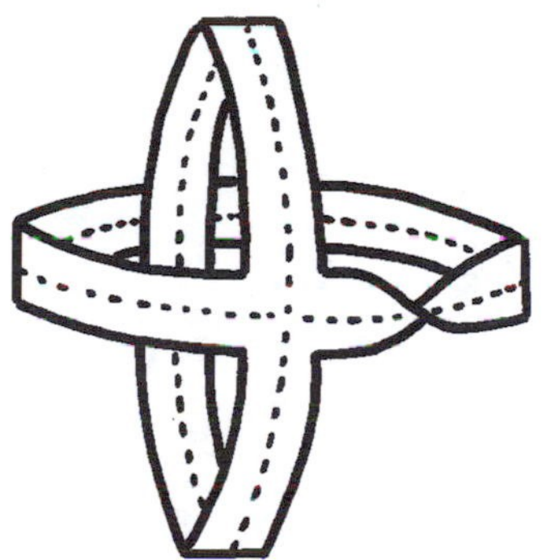

9

The Infamous Bicycle Problem

9.1 Which Way Did the Bicycle Go?

Here's a problem gaining some notoriety.

These tracks were left by a bicycle whose wheels were colored with sidewalk chalk. Which way did the bicycle go, and what was the length of the bicycle?

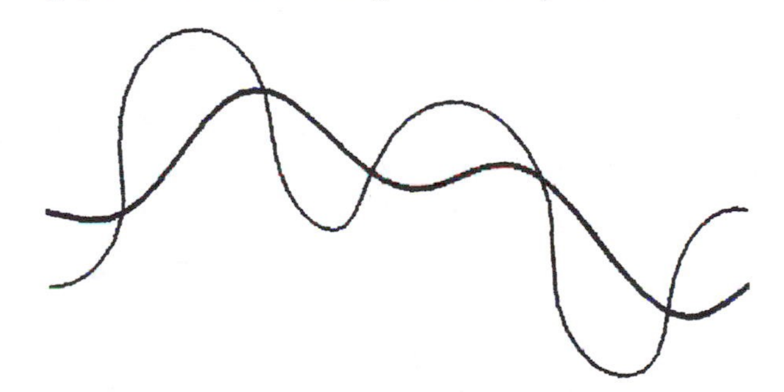

Dortheanne Roberts rides along a 15-foot length of paper. The wheels were colored with sidewalk chalk and the tracks left on the paper were later traced over with permanent markers to make them bolder.

9.2 Pedal Power

A bicycle is held gently by the seat to keep it balanced while the pedal situated at its lowest position is pushed backwards. Which way does the bicycle move? Forward or backward?

Josh Davis holds the bicycle while Diane Dixon pushes on the pedal.

9.3 Yo-Yo Quirk

A yo-yo, with its string wound around the spool in the direction shown, is placed on its edge on a table top. If the string is gently pulled, which way will the yo-yo move: forward with the pull or backward with the unwinding of the string?

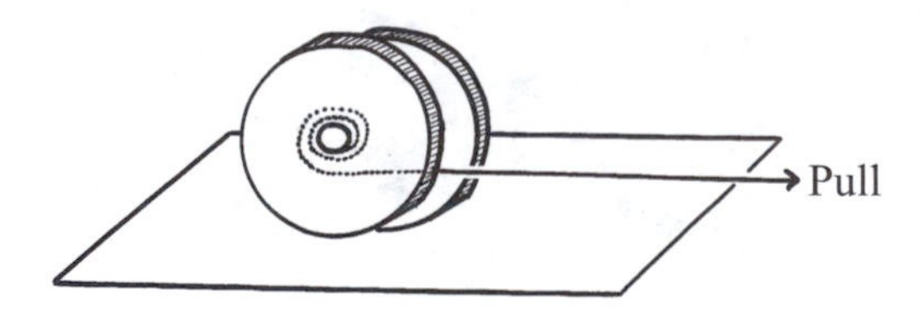

10

Making Surfaces in 3- and 4-Dimensional Space

10.1 Making a Torus

To form a *torus* (donut shape) from a square piece of paper, simply glue the top edge to the bottom edge to form a cylinder and then bend this cylinder to glue the left edge to the right edge. (Topologists say that the opposite edges have thus been *identified*.)

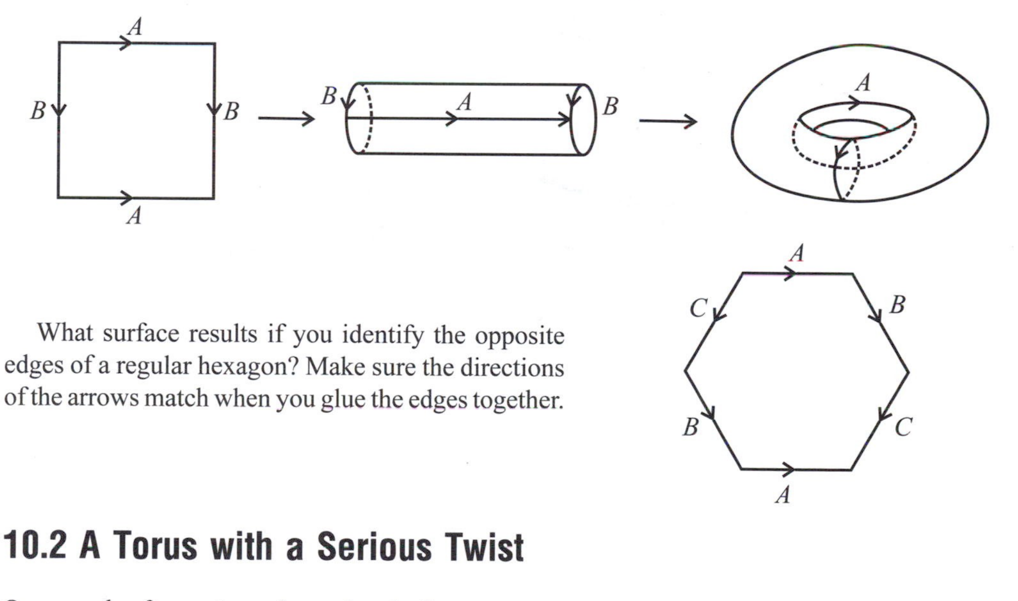

What surface results if you identify the opposite edges of a regular hexagon? Make sure the directions of the arrows match when you glue the edges together.

10.2 A Torus with a Serious Twist

One can also form a torus from a band of paper by gluing the top edge to the bottom edge all the way around. (In practice, however, this is tricky to do. Paper is very unforgiving and won't stretch. The donut you obtain, after much perseverance, will be quite crumpled!)

Suppose this band is given a half twist in its construction. Can we still obtain a torus by gluing the top edge to the bottom edge?

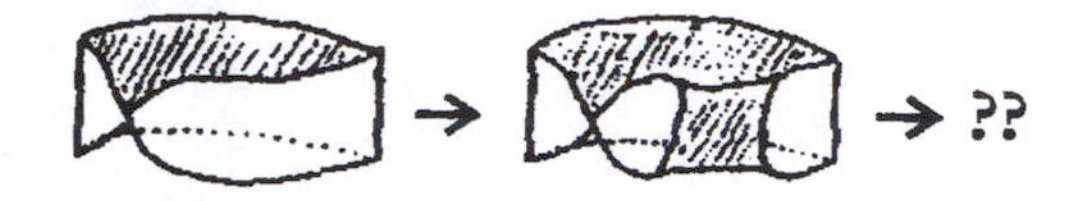

Comment. Try this exercise with paper and tape, fabric, needle and thread, or better yet, Play-Doh®.

10.3 Capping Möbius

The boundary edge of a Möbius band is simply a circle: one that has been bent into a squashed figure eight.

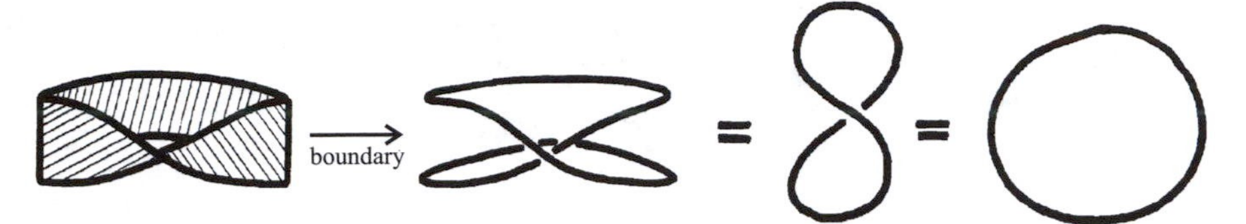

The boundary of a circular disc is also a circle. What happens if we sew these two figures together along their common boundary circles? What surface results?

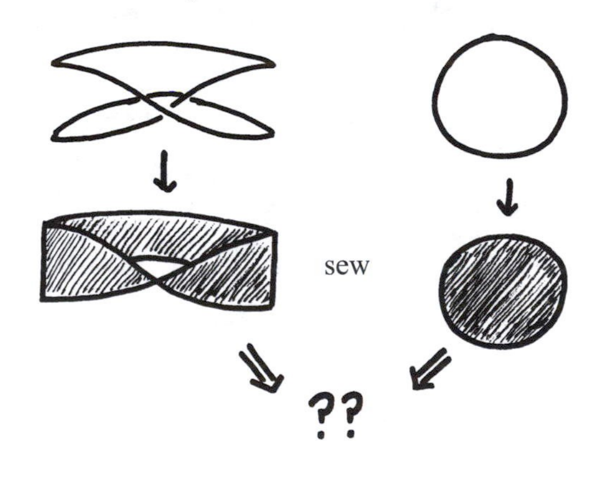

11

Paradoxes in Probability Theory

11.1 The Money or the Goat?

This classic puzzle from probability theory is known as the Monty Hall problem after the host of "Let's Make a Deal!" It was played as part of that TV game show.

Imagine you are a game show contestant with three closed doors before you. You are told that behind one of these doors is a prize of one million dollars in cash. Behind each of the other two are goats. You select a door, but before you open it, the host opens one of the two remaining doors to reveal—a goat! He now gives you the opportunity to stay with your original choice or switch to the remaining unopened door. What should you do? Switch doors or stay with your initial choice? Does it make any difference?

11.2 Double or ... Double!!

A friend places before you two paper bags, both containing Tootsie Rolls®. He tells you that one contains twice as many as the other and that you may keep the contents of one of the bags. You select a bag, open it up, and count the number of Tootsie Rolls® it contains. Your friend then gives you the option to change your mind and take instead (without peeking inside!) the contents of the other bag. Assuming you would like as many Tootsie Rolls® as possible (and you don't feel it is an insult to his generosity to switch), is it to your advantage to switch bags? Or is it better to stay with your first choice?

11.3 Discord Among the Chords

This puzzler is known as Bertrand's Paradox.

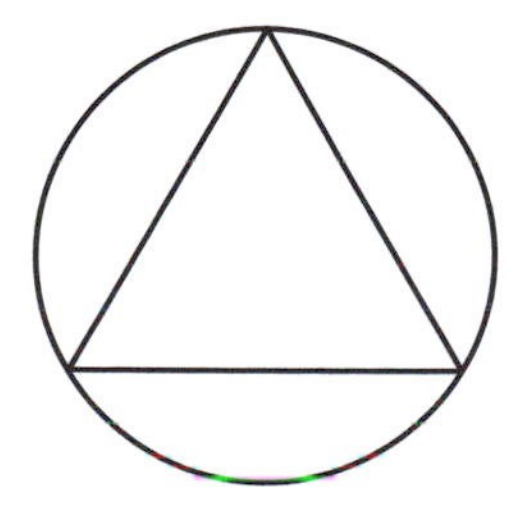

Consider a circle of radius R. Inside this circle inscribe an equilateral triangle. This triangle has side length $\sqrt{3}R$. Suppose a chord of the circle is selected at random. What is the probability P that the length of this chord is greater than $\sqrt{3}R$, the side length of the triangle?

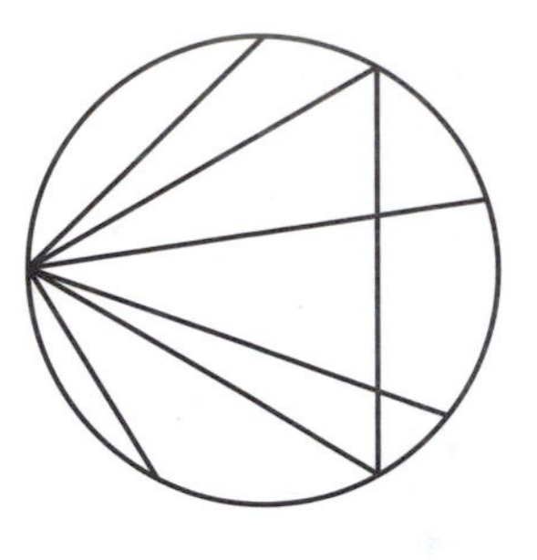

Jennifer answers this question in the following way: Once a chord is drawn, we can always rotate the picture of the circle so that one end of the selected chord is placed at the leftmost position of the circle. We may as well assume then that all chords considered in this problem have one end point at this left-most position. Now draw the equilateral triangle as shown. It is clear that the length of the chord will be greater than the side length of the triangle if the other end point lies in the middle third of the perimeter of the circle. Thus $P = \frac{1}{3}$.

Bill reasons: If the midpoint of the chord lies anywhere in the shaded region shown below, its length will be greater than the side length of the triangle. Thus P equals the probability that the midpoint lies in this region. A quick calculation shows that the area of this region is one quarter of the area of the entire circle, thus $P = \frac{1}{4}$.

Joi, on the other hand, argues this way: Rotating the picture of the circle and the selected chord, we may assume that the chord chosen is horizontal. If the midpoint of this chord lies on the solid segment of the vertical line shown, its length will be greater than the side length of the triangle. Thus $P = \frac{1}{2}$.

Whose reasoning is correct?

Taking It Further. What would happen if you threw a handful of wires onto a circle drawn on the ground and measured the lengths of the chords crossing that circle? Approximately how many would be of length greater than $\sqrt{3}R$?

11.4 Alternative Dice

An ordinary six-sided die has faces labelled 1 through 6. Thus, given a pair of dice, the probability of rolling a sum of 12 is $\frac{1}{36}$; a sum of 8 is $\frac{5}{36}$; a sum of 4 is $\frac{3}{36}$; and so on. Is it possible to relabel the faces of two six-sided dice with alternative positive integers so as to produce two dice with the same sum probabilities as ordinary dice?

Taking It Further. Two "ordinary" tetrahedral dice have sides labelled 1, 2, 3, and 4. Is there a clever relabelling that yields the same sum probabilities as the ordinary dice?

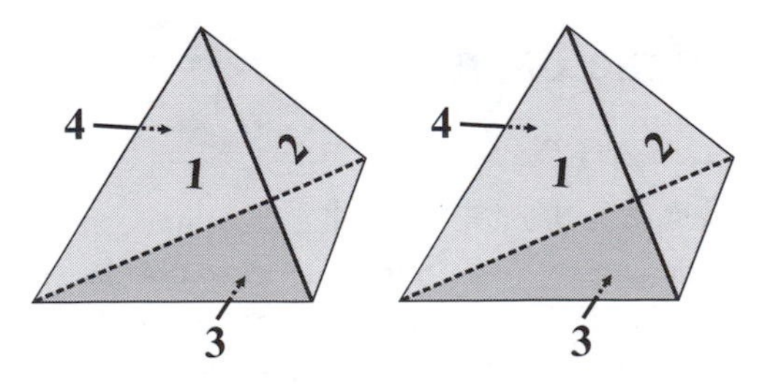

12

Don't Turn Around Just Once

12.1 Teacup Twists

Andy holds up a teacup in the center of a room while his friends tape several strings from the teacup to various points about the room, leaving plenty of slack for later maneuverability. Next, Andy carefully rotates the cup 360°, tangling the strings in the process, and then holds the cup

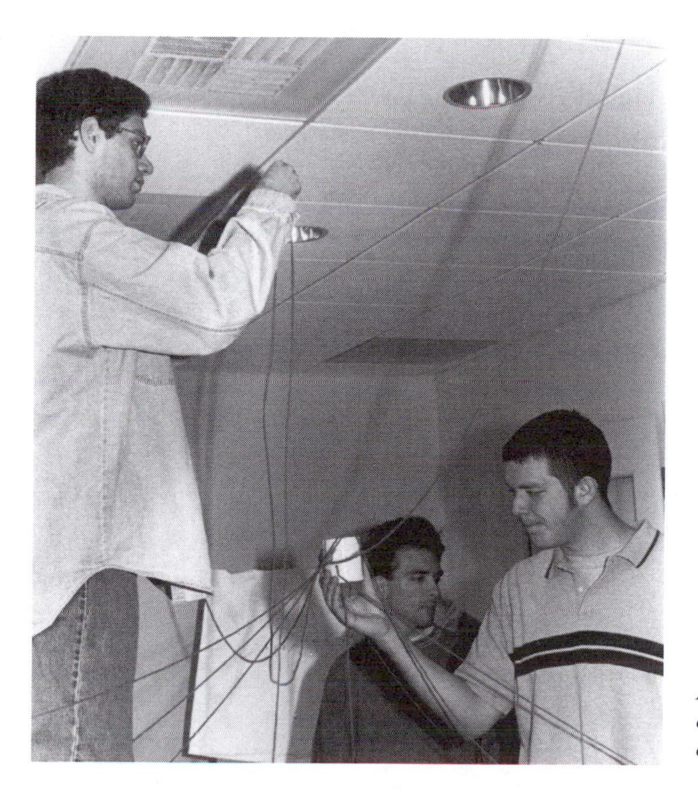

Andy Furey holds a teacup while club members attach strings to various points around the room.

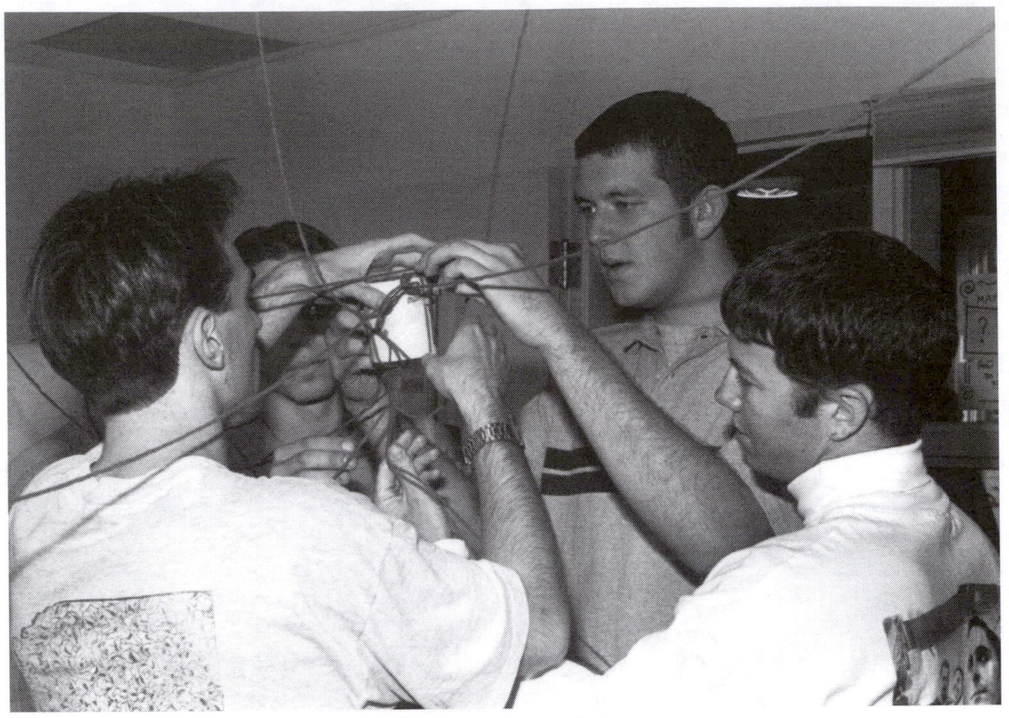

Andy and the tangled cup.

firmly at that point in space never to be moved again! Is it possible for his friends to maneuver the strings around the teacup and untangle them?

Later they decide to try the experiment again. This time Andy gives the cup two full turns, 720°, tangling the strings even more than before! In this situation, is it possible for his friends to maneuver the strings around the teacup, again held in place, and untangle them?

12.2 Rubber Bands and Pencils

Wrap a rubber band around the end of a pencil so that the band always lies flat against the wood. How many times *must* the band wrap around the pencil to achieve this?

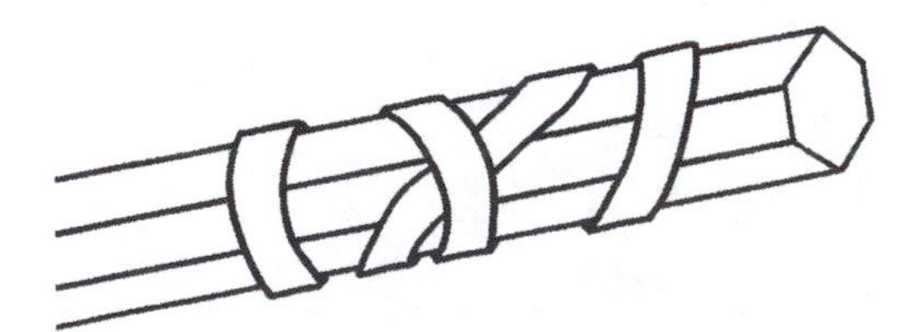

13

It's All in a Square

13.1 Square Maneuvers

Twenty-five people stand in a large 5×5 square grid, one person per cell. Each person is asked to take one step (vertical or horizontal, but not diagonal) into a neighboring cell, that is, a cell sharing one entire edge with his or her current cell. People are allowed to exchange squares, but no one may share a square. Is it possible to end up with a new arrangement of all 25 in the square grid?

Taking It Further. Suppose everyone is energetic and decides instead to all leap to a square *two* places away, either in a vertical or horizontal direction. Is the puzzle solvable?

Elbows fly as 25 St. Mary's College students seek new squares.

13.2 Path Walking

Starting at the top left corner of a 7×7 square grid, it is possible to walk a path, using vertical and horizontal motions only, that visits each and every cell of the grid precisely once. Is such a path possible starting one square over from the top left one? From which cells is it possible to commence such a path?

Taking It Further. Is it ever possible, when walking a path, to return to your initial cell and form a loop of steps that visits each and every cell precisely once?

Comment. To get a feel for this problem try walking in the large 5×5 grid you drew in section 13.1.

13.3 Square Folding

Construct a 4×4 square grid and label the cells as shown. Repeatedly fold the grid along its straight line markings to reduce it to a 1×1 wad of paper 16 layers thick. Be as devilish with your folding as you like, sneakily tucking folds within each other, turning the square over multiple times in the process, and the like. Once done, trim away the edges of this unit square, and without disturbing the orientation of the 16 square layers, spread them out over the table. Sum all the numbers you see. What do you get?

4	12	10	4
5	2	9	13
4	11	11	8
7	16	7	3

Repeat this experiment several times or have several people do it at the same time. What do you notice? (You are in for a surprise!)

14

Bagel Math

14.1 Slicing a Bagel

One normally cuts a bagel with a horizontal planar slice. The image of this cut on the slicing plane consists of two perfect circles bounding a region of dough.

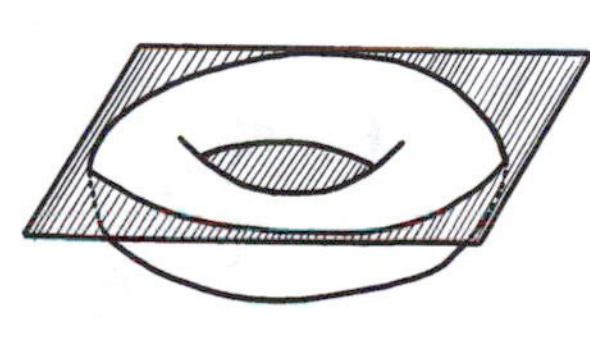

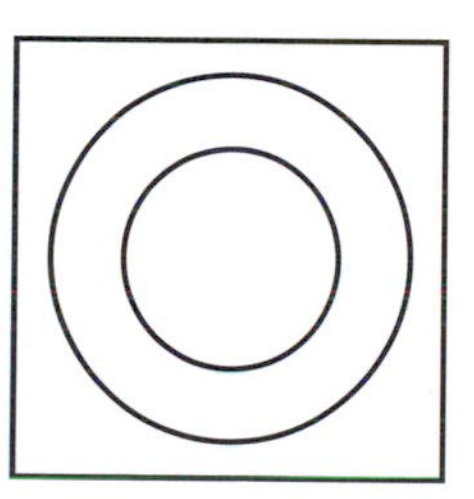

When sharing a bagel with a friend you might cut the bagel in half through a vertical plane. This also produces an image of two perfect circles on the slicing plane.

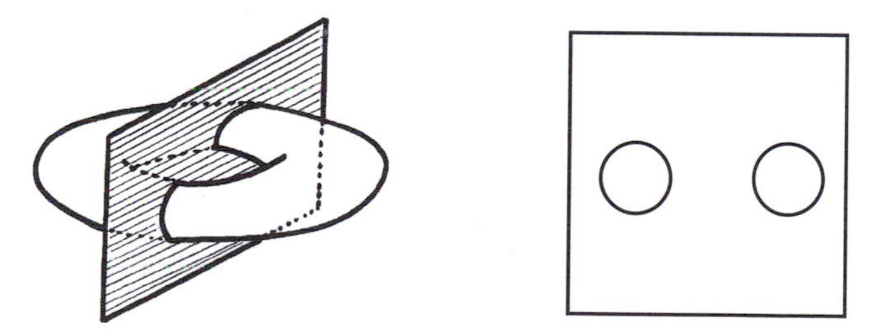

How else could one slice a bagel so as to produce the image of two perfect circles on the slicing plane? Assume we are working with perfectly toroidal bagels!

14.2 Disproving the Obvious

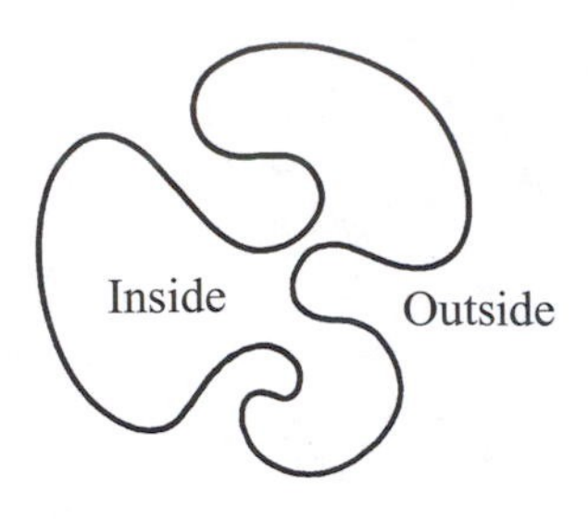

It seems completely and utterly obvious that a closed loop divides whatever it is drawn on into two distinct pieces: an inside and an outside. Show that this "utterly obvious" theorem is in fact false for some curves drawn on the surface of a bagel!

14.3 Housing on a Bagel

Here's a classic problem from graph theory.

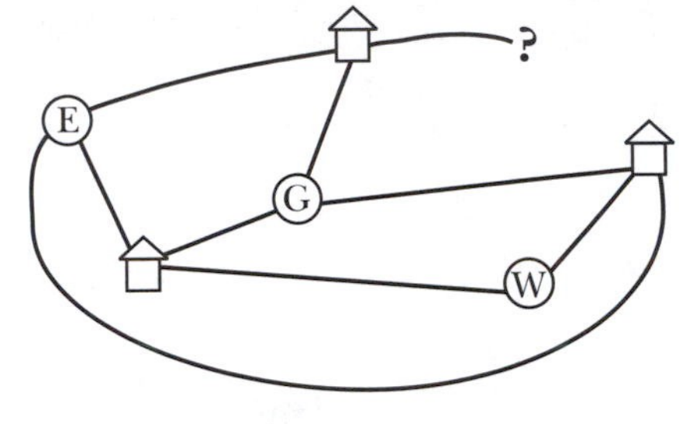

Three houses must be connected to three utility companies (for electricity, water and gas) in such a way that no lines or mains cross. Is this possible?

Surprisingly, for buildings situated on a plane this problem is never solvable. (Try it!) However, on our spherical earth we have the (theoretical) option of allowing pipes to circumnavigate the entire globe. Is this problem solvable on a sphere?

14.4 Tricky Triangulations

Here I have covered a bagel with 64 warped triangular regions in such a way that any two neighboring triangles touch either in a *single* vertex or along just *one entire* edge. Could I have accomplished this feat with an *odd* number of triangles? I could have done this with less than 64 triangles. Show that it is possible to cover a bagel with just 14 triangles but no fewer!

14.5 Platonic Bagels

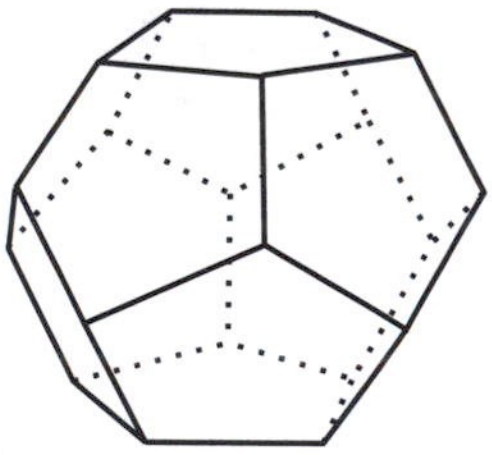

A dodecahedron, the fourth platonic solid, can be formed by sewing together 12 pentagons. Notice that the same number of edges meet at each vertex.

Is it possible to cover a bagel with five-sided regions so that the same number of edges meet at each vertex?

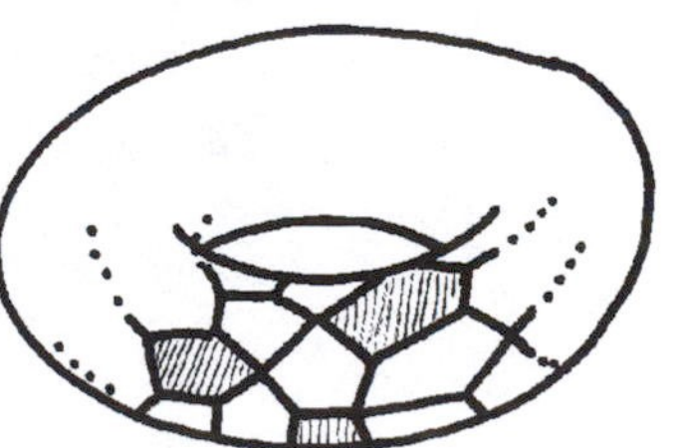

15

Capturing Chaos

15.1 Feedback Frenzy

Most video cameras can be hooked up to a TV to display what the video camera sees in real time. Thus if you point the video camera at the TV screen (being careful to exclude the sides of the TV) the screen displays an image of the screen itself. The image is blank. This remains so even if you focus on only a small portion of the screen or if you hold the camera at a tilt.

Now hold the camera at an angle to horizontal. Dim the lights in the room and place a lit candle between the camera and the TV. The camera will see the candle and display its image on the screen, which it then sees and displays again, which again it sees and again displays, and so on. What appears is a beautiful swirling image of a flame dancing around the screen. (Try it!)

If you experiment carefully with the camera and TV alignment, you can obtain a situation where the image of the candle does not disappear even if you blow out the flame. At this point you have captured the image on screen. Since the flame and its slight motion are no longer present, you

Lane Anderson of St. Mary's College points a video camera at a screen showing its own output.

would expect a stable image, frozen in time, to be left on the screen, but this is not the case! Instead the spectacular swirling dance continues, forever captured on the screen! What is going on?

Comment. This is quite finicky to set up and much perseverance is required. Place the camera on a tripod and turn off all automatic features. Adjust the zoom on the camera until the image of the screen is almost the size of the screen itself. Set the TV brightness on low and then light the candle. Adjust the color, focus, zoom, and brightness until interesting effects occur. You may even want to place tape with a small pinpoint hole over the lens of the camera.

15.2 Creeping up on Chaos

Setting $a_0=0.1$ and $r=2$, consider the *recursive relation*

$$a_{n+1} = ra_n(1 - a_n).$$

This defines a sequence of values

$$a_1 = ra_0(1-a_0) = 0.180$$
$$a_2 = ra_1(1-a_1) = 0.295$$
$$\vdots$$

Using a calculator or computer we can easily determine the first ten terms of this sequence:

a_1	a_2	a_3	a_4	a_5	a_6	a_7	a_8	a_9	a_{10}
0.180	0.295	0.416	0.486	0.500	0.500	0.500	0.500	0.500	0.500

The sequence appears to converge to the value 0.500.

The same type of behavior occurs if we repeat the exercise with the value $r = 2.5$, though the limit of the sequence appears to be different.

a_1	a_2	a_3	a_4	a_5	a_6	a_7	a_8	a_9	a_{10}
0.225	0.436	0.615	0.592	0.604	0.598	0.601	0.600	0.600	0.600

However, curious things occur if you repeat this exercise with higher values of r. For instance, set $r = 3.3$. If you have the patience, calculate the first 20 terms of the sequence. What do you notice? Now set $r = 3.4$ and then 3.5 and calculate the first 20 terms of each sequence. What do you observe? Check out what's happening for $r = 3.54$ and $r = 3.55$. (You may need to calculate at least 30 terms of the sequence here.) Finally, given the first 20 terms of the sequence for $r = 3.70$, can you predict what the 21st term will be?

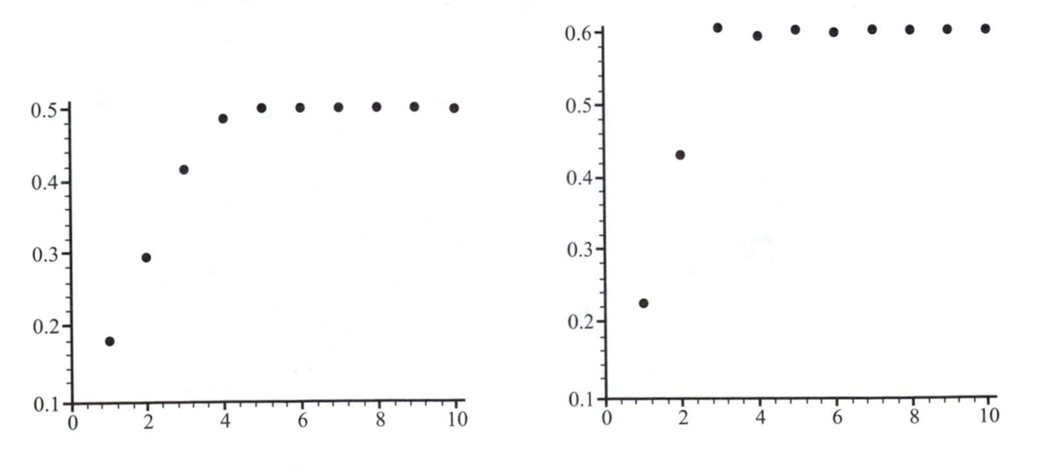

16

Who has the Advantage?

16.1 A Fair Game?

Peter has ten coins, Pennelope has nine. Peter and Pennelope agree to toss all their coins simultaneously. Whoever receives the largest number of heads will win. In case of a tie Pennelope will be declared the winner, so as to offset the advantage Peter has to begin with. Given this agreement, who is most likely to win?

Find a partner and experiment with this game a few times. Which player appears to be favored?

16.2 Voting for Pizza

Alice, Brad, and Cassandra decide to order a pizza to share. Alice will pay for the pizza but she only has enough money to order one topping: pepperoni, anchovies, or olives. Alice prefers pepperoni, is indifferent about anchovies, but simply detests olives. Brad and Cassandra have equally strong opinions, Brad preferring anchovies to olives, and olives to pepperoni, and Cassandra olives to pepperoni, and pepperoni to anchovies.

	Alice	Brad	Cassandra
1	Pepperoni	Anchovies	Olives
2	Anchovies	Olives	Pepperoni
3	Olives	Pepperoni	Anchovies

Seeing no clear group preference, they decide to vote. They will write their choice of topping on a slip of paper, and the topping listed the greatest number of times among the ballots will be the one chosen. In case of a three-way tie, pepperoni will prevail. This voting advantage is given to Alice, who is paying for the pizza, after all. If each person is a savvy player, what topping will the group end up ordering? Play this game with a group of three and see what happens.

John Dockstader, Angela Dellano, and Kelly Ogden of Merrimack College engage in a three-way duel with dice.

16.3 A Three Way Duel

Here's a classic puzzle from probability theory with a counterintuitive conclusion.

Three people, armed with pistols but unequal in marksmanship, enter into a three-way "duel." Alberto can hit his target on average one third of the time. Bridget, on average, hits her target two thirds of the time, but Case is a perfect shooter — he always hits his target.

Being gentle folk, these participants agree on a shooting order that reflects their shooting abilities. Alberto will shoot first, aiming in any direction he desires, then Bridget will shoot (if she is still alive), to be followed by Case (if alive), then back to Alberto, and so on in cyclic order until just one person is left standing. What is the optimal strategy for each player?

For Case, clearly his best strategy is always to shoot the more competent opponent remaining — namely Bridget, if she is still alive. Thus Bridget should always aim for Case, knowing she is his prime target. But what should Alberto do? Should Alberto follow the same strategy and aim for the most competent player alive? What are his chances of survival?

Comment. Use dice for guns, where rolling a 1 or a 2 is a successful shot for Alberto, a 1, 2, 3, or 4 is a successful shot for Bridget. Play out this game a large number of times and compute the average survival rate for each player. Have everyone always aim for the more competent opponent alive.

16.4 Weird Dice

Consider the unusual numbering schemes illustrated on these dice. You pick one die, and then I will pick another. We each will roll our chosen die and the larger number wins. Do you want to play with me?

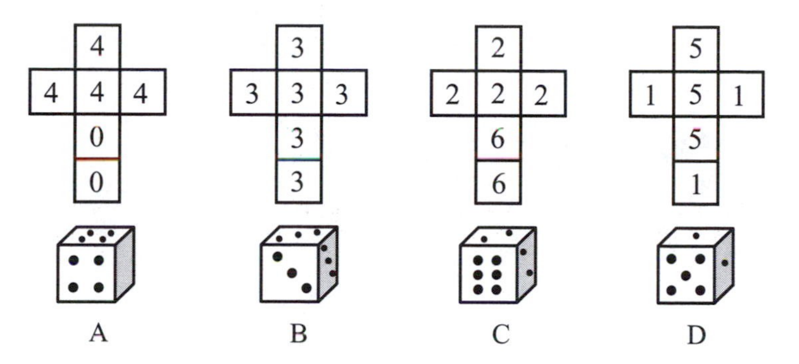

Comment. Try experimenting with dice made of colored paper.

17

Laundry Math

17.1 Turning Clothes Inside Out

A T-shirt inside out looks the same in shape and structure as a T-shirt right-side out; only the seamwork and the patterns on the material tell you something is amiss. The same thing is true for socks, trousers, skirts, and jackets. Is this always the case? Does the process of *eversion*, turning things inside out, preserve the general structure and shape of all clothes? If so, why?

Aside. For a while it was fashionable to deliberately wear sweatshirts inside out. In taking off a sweatshirt do you grab the neck opening and pull it over your head, preserving the orientation of the shirt throughout the process, or do you grab from the waist and turn the shirt inside out as you take it off? Do you take note of the effect when you next put it on? What does this reveal about you?

Dennis Horton attempts to turn a pair of mutilated trousers inside out.

17.2 Mutilated Laundry

Let's attack our laundry with needle and thread.

Sewing together the two leg openings of a pair of trousers yields a punctured donut.

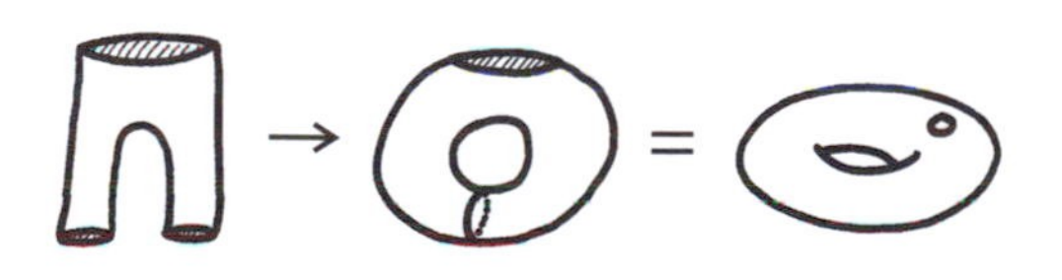

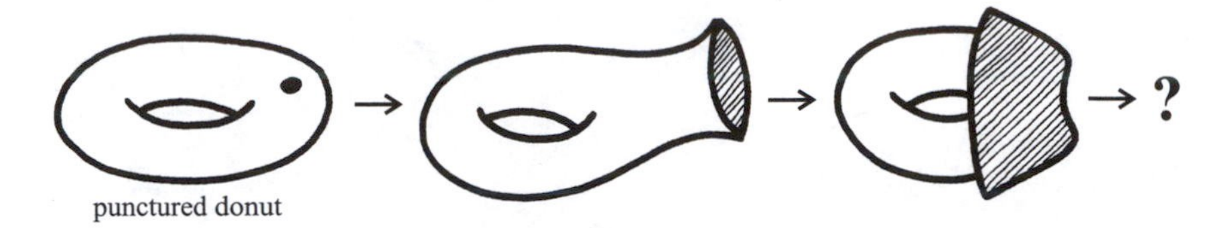

What happens if you turn a punctured donut inside out? Is this possible? If so, what shape do you obtain?

A jumpsuit (that is, a one-piece shirt and trouser outfit) provides a means for creating a *double-donut*. Using sufficiently bizarre items of clothing (or sewing multiple pieces of clothing together) we can also create *triple-donuts*, *quadruple-donuts* and other *multi-donuts*.

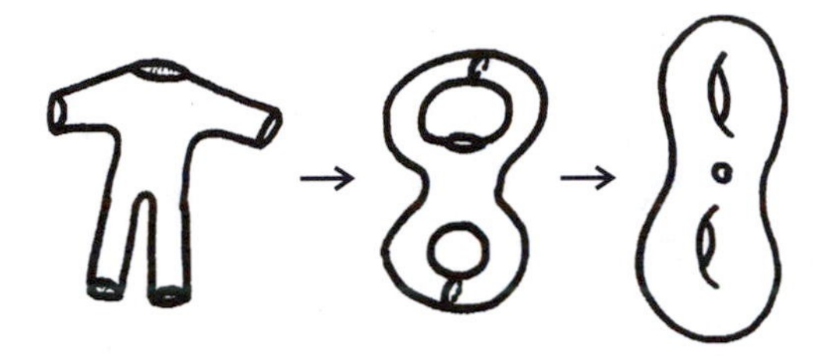

What happens if you turn a punctured double-donut inside out? An arbitrary multi-donut?

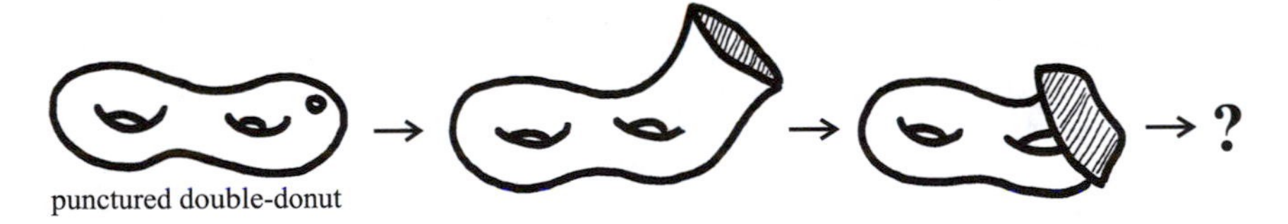

Comment. Stapling is quicker and easier than sewing.

17.3 Cannabalistic Clothing

Take two pairs of trousers and connect them to form two linked donuts. Open up the puncture of one donut (as in eversion) and "swallow" the second. What happens? Is the result complicated or elegantly simple? How exactly does the second donut sit inside the stomach of the first?

18

Get Knotted

18.1 Party Trick I: Two Linked Rings?

One end of a long piece of string is tied around Jason's left wrist, and the other end is tied around his right wrist. Paul's wrists are also tied, but his string passes through the loop created by Jason's arms and his string. Can these two gentlemen separate themselves from their linked dilemma? They may move the string any way they like, step through any loops, or wrap around themselves in any clever way.

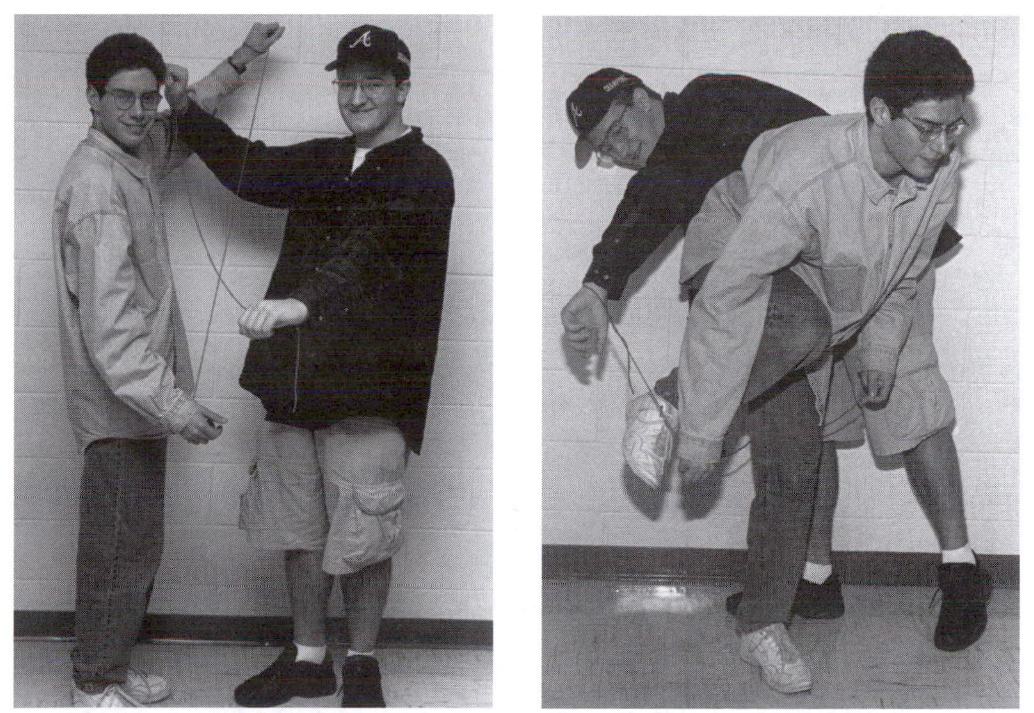

Jason Summers and Paul Ogle are trapped as two linked rings.

Comment. Tying the strings to their wrists is easier than simply holding the strings in their hands. This leaves their hands free for complex maneuvering.

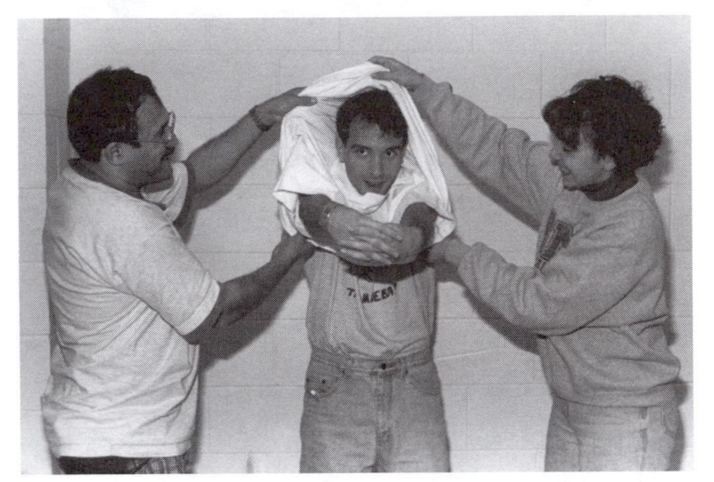

Moncef Boufaida and Dortheanne Roberts try to turn George Hinkel's shirt inside out.

18.2 Party Trick II: A T-Shirt Trick

George, wearing a baggy T-shirt over his clothes, holds his clasped hands out in front of him. Is it possible for his friends to remove his T-shirt, turn it inside out, and put it back on him while George *keeps his hands firmly clasped together*?

Taking It Further. George is again wearing the baggy T-shirt over his clothes and Aliza is firmly holding each of his wrists. Is it possible to remove the T-shirt from George and place it on Aliza all the while Aliza keeps her grip?

Aliza Steurer holds George Hinkel's wrists.

Sten-Ove Uva wears a waistcoat and jacket. Josh Davis and Lusine Ayrapetian try to turn the waistcoat inside out.

18.3 Party Trick III: A Waistcoat Trick

Sten-Ove is wearing a baggy waistcoat, unbuttoned, underneath his jacket, also unbuttoned. Is it possible for Josh to take off Sten-Ove's waistcoat, turn it inside out, and place it back on underneath the jacket *without* slipping any material down inside the sleeves of the jacket?

18.4 Two More Linked Rings?

John can touch the tip of each thumb with his index fingers to form two unlinked rings or two linked rings. If John were composed entirely of soft clay and if we only stretch and mold the clay (no puncturing or tearing of material allowed: keep neighboring molecules of clay neighbors)

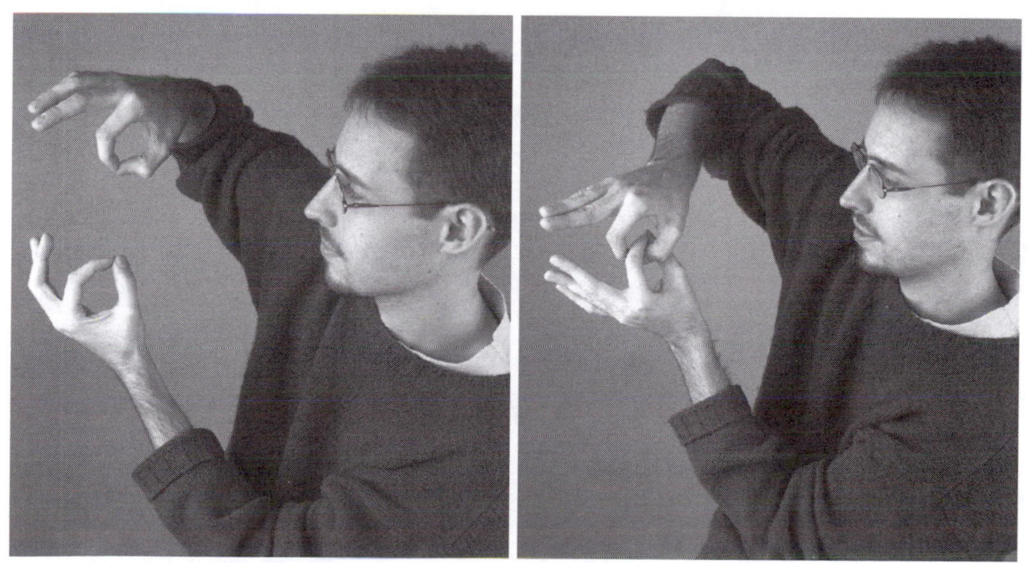

John Dockstader forms rings with his thumbs and index fingers.

can we smoothly transform the picture on the left to that on the right? Are these two linked rings, in some sense, no different from two unlinked rings?

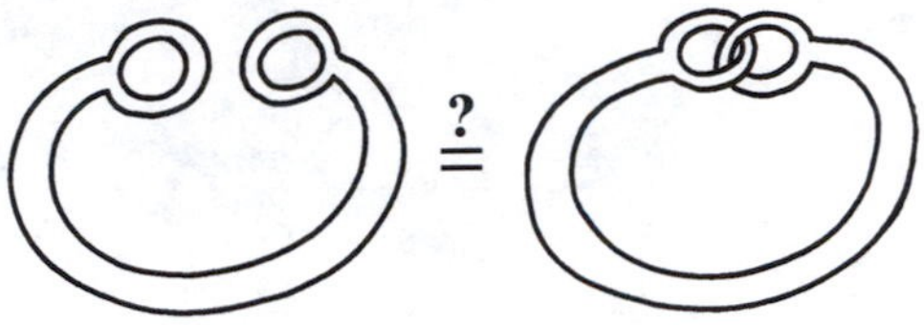

$\stackrel{?}{=}$

19

Tiling and Walking

19.1 Skew Tetrominoes

A *tetromino* is any tile composed of four connected cells from a square lattice (a *domino*, on the other hand, is a tile composed of just two). There are 19 tetrominoes, each being a reflection or rotation of one of five basic configurations, called types I, II, III, IV, and V. Those of type V are called the *skew tetrominoes.*

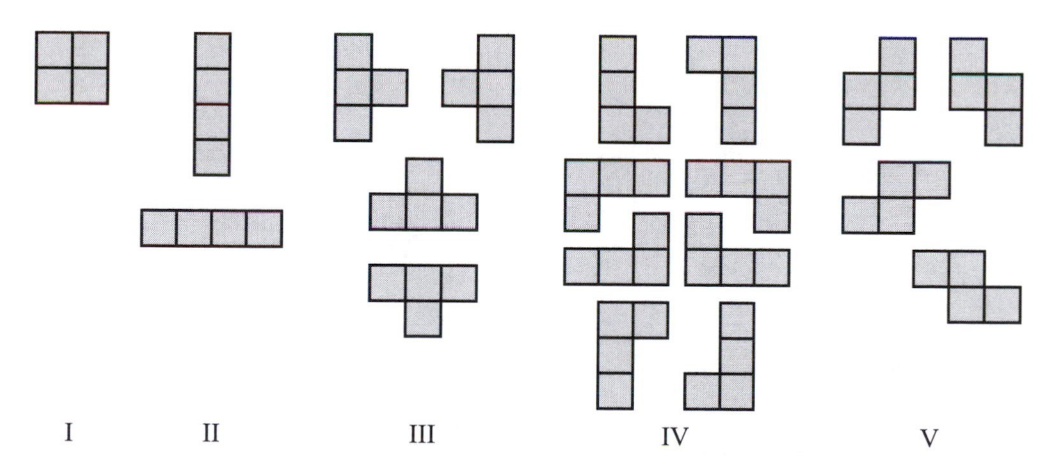

It is possible to tile a 4×4 square with four tiles of type I, four of type II, four of type III, or four of type IV. No four skew tetrominoes, however, will tile this square. Is there a square or rectangular grid of any size that can be completely tiled with non-overlapping skew tetrominoes?

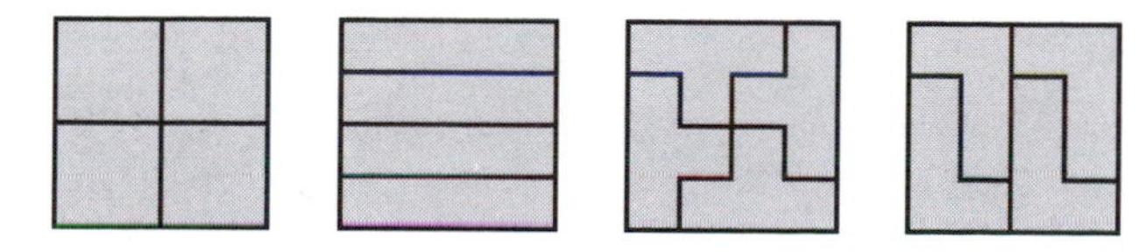

19.2 Map Walking

The diagrams below represent the location of streets and avenues in Adelaide (city A) and Brisbane (city B). The thoroughfares divide each downtown area into square city blocks, with the streets running east to west and the avenues north to south. Traffic is allowed to move only in the directions indicated. (Do the inhabitants of city A have a problem?)

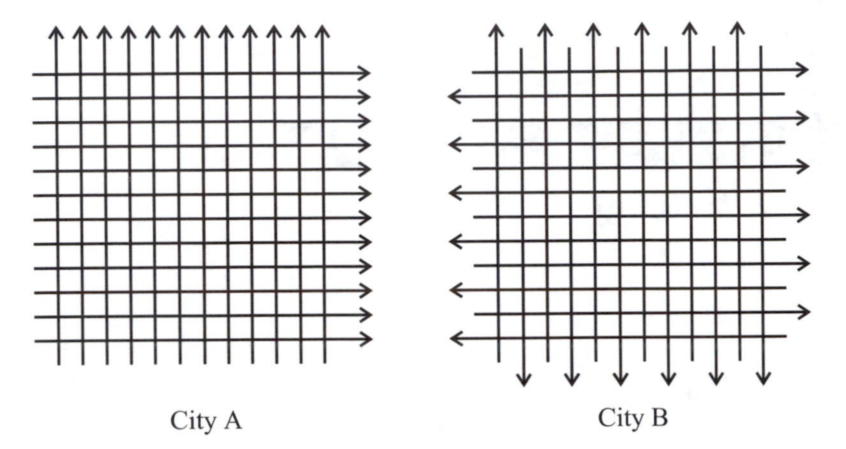

City A City B

Point to an intersection in city A and have a friend point to one in city B. Imagine, as a pedestrian, you are walking through city A, following the streets and avenues from intersection to intersection, going either with or against the traffic along the roads. Describe your journey to your friend simply by stating whether you are moving along a street or an avenue, with or against the traffic, as you move along the city blocks. For example, the path illustrated below takes you back to your starting point and would be described as

$$SSAS^{-1}AAS^{-1}S^{-1}A^{-1}A^{-1}SA^{-1}$$

where S means "move one block along a street with the traffic," A^{-1} means "move one block along an avenue against the traffic," and so on.

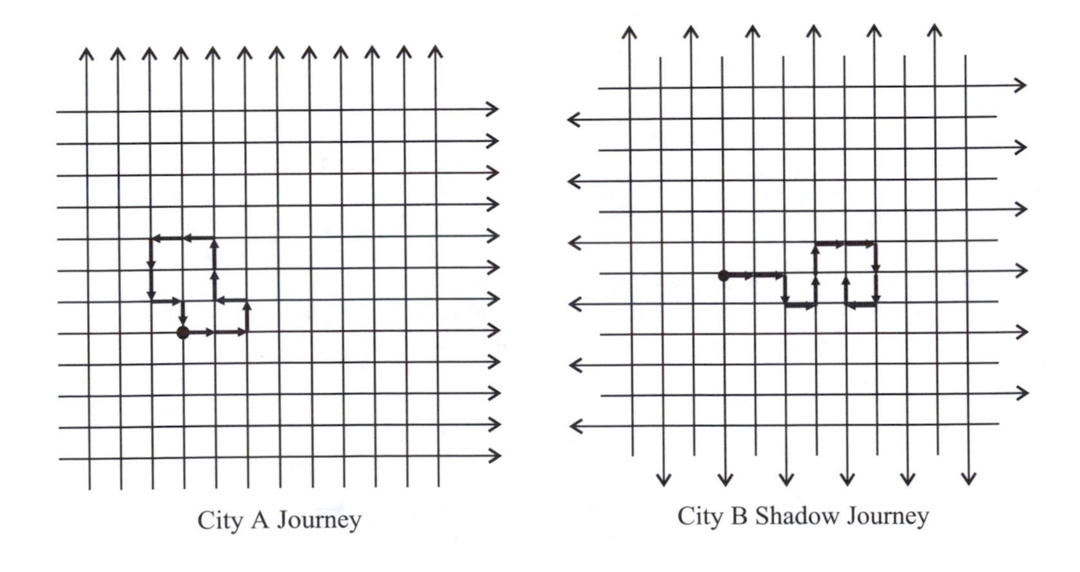

City A Journey City B Shadow Journey

Your friend, hearing the description of your journey, will follow suit on her City B map and walk a "shadow journey," moving along streets and avenues, with or against the traffic, as per your instructions. In the example presented here, your friend's journey is very different from your own: it does not even form a closed loop, for example.

Does there exist a closed loop journey in city A that results in a closed loop shadow journey in city B? If you find one, does it depend on the particular intersection at which your friend starts her journey?

19.3 Bringing It Together

Can this region be tiled with skew tetrominoes?

Hint. Consider the title of this section!

20

Automata Antics

20.1 Basic Ant Walking

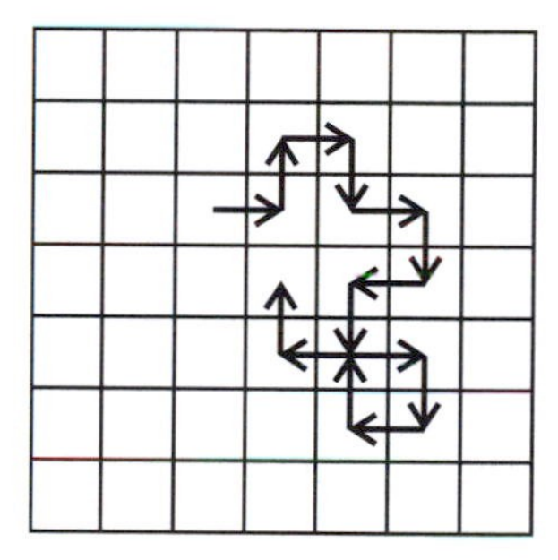

An ant moves about a 7×7 grid of squares, taking single steps in alternating vertical and horizontal directions. If the ant enters a cell from a horizontal direction, can it ever visit that cell again from a vertical direction?

20.2 Ant Antics

An ant moves about a 7×7 square grid beginning in the central square, facing north. Each cell is labelled L for left or R for right. The ant moves according to a set procedure: It takes a step forward and looks at the label of its new cell. If the cell is labelled L the ant turns left 90°, right 90° if it is labelled R. The ant then changes the label of the cell (from R to L, or L to R), takes its next step forward, and repeats this procedure over and over again. Given these rules of motion, is it possible to devise an initial labelling scheme of the cells so that the ant is *not* forced to leave the 7×7 grid?

R	L	L	R	R	L	R
R	L	R	L	R	L	L
L	R	L	L	R	R	R
L	L	R	L	L	R	L
R	R	L	R	R	L	R
R	L	R	L	L	L	R
L	L	R	R	L	R	L

20.3 Ball Throwing

A number of students stand in a circle, each with the word left or right in mind. John begins a ball game by tossing a ball across the circle. If Beatrice catches the ball she throws it back across the circle one place to the *left* (in her perspective) of John if she is thinking left, or one place to the *right* if thinking right. Then Beatrice switches words ("right" becomes "left" and vice versa) and waits for another turn. Each person receiving the ball operates in this way.

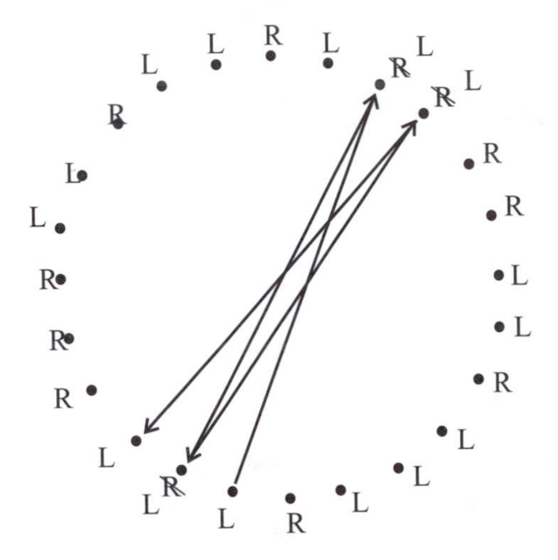

There is one complication: If Maxine is thinking "right" and receives the ball from Christian directly to her right, the rules do not make sense. We make the convention that in this predicament Maxine holds the ball, turns in place to the right more than 180°, and throws the ball to the first person she sees (and still changes the word she holds in mind). This is, of course, equivalent to Maxine passing the ball to the person directly on her left.

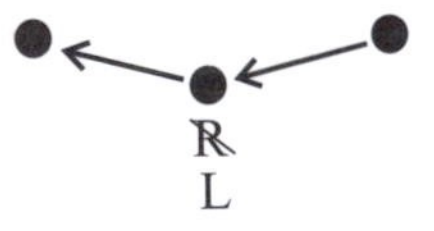

Similarly, if Keone is thinking "left" and receives the ball from Shannon directly to his left, he turns left and throws the ball to the person on his right. This is a little confusing at first, but it doesn't take long to get the hang of it.

Here is the question: In this game is everyone guaranteed a turn? Will the ball eventually reach everyone, no matter the choice of word held initially in everyone's mind?

Comment. I recommend a lightweight ball if playing this game indoors!

21

Bubble Trouble

21.1 Road Building

Here's a classic puzzle.

Four towns, situated on a plane at the vertices of a square, are to be connected by a road system using the minimum total length of road. Costs of construction are of importance here, not the convenience of the towns' inhabitants. Should the local government settle on one of the designs below? If so, which one? Is there an even better solution?

21.2 Higher Dimensional "Road Building"

Let's take the problem up a dimension.

What design of surfaces, meeting somewhere in the center, connects the skeleton of a cube (namely its 12 edges and 8 vertices)with minimal total surface area?

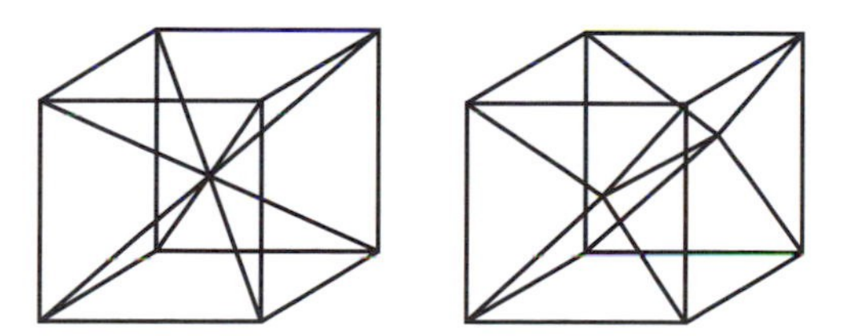

This problem is very difficult to analyze mathematically, but with the aid of soap solution the answer can be determined experimentally. Using pliable wire make a frame of a cube and dip it into soap solution. The surface tension of the liquid film acts to minimize surface area, so careful dipping (making sure a film is attached to every edge of the cube) will result in the desired solution to the problem.

What happens if you gently shake the structure of film you obtain? How does the solution to the problem change?

21.3 Donut Bubbles

Is it possible to make a stable donut-shaped bubble? (Try it!)

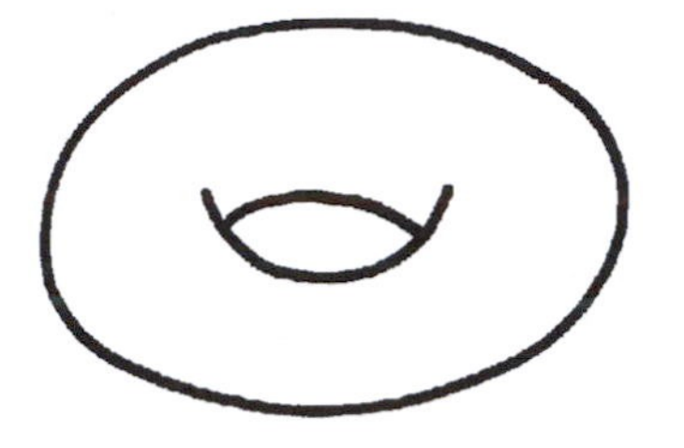

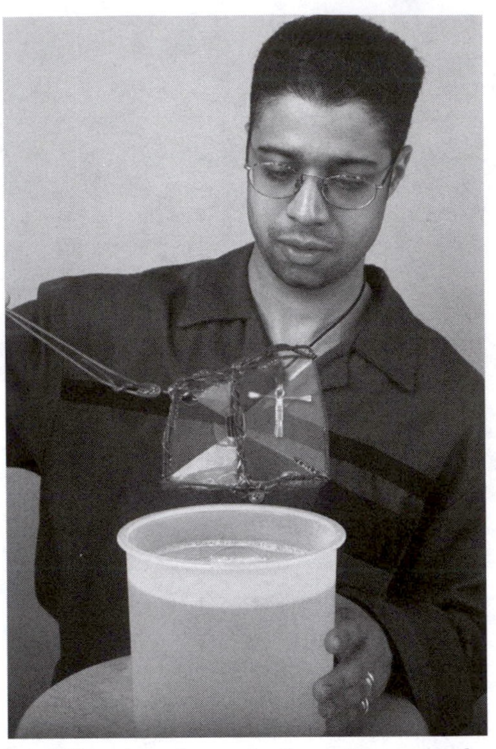

Frank Francisco dips a wire frame into soap solution.

Making wire frames and soap solution. Florist's wire, or 18 gauge aluminum wire works well. Toy stores sell bubble solution in large containers. Or you can make your own: Mix together one gallon of hot water, one cup of Dawn liquid detergent, and one tablespoon of glycerin; let the mixture sit overnight.

22

Halves and Doubles

22.1 Freaky Wheels I

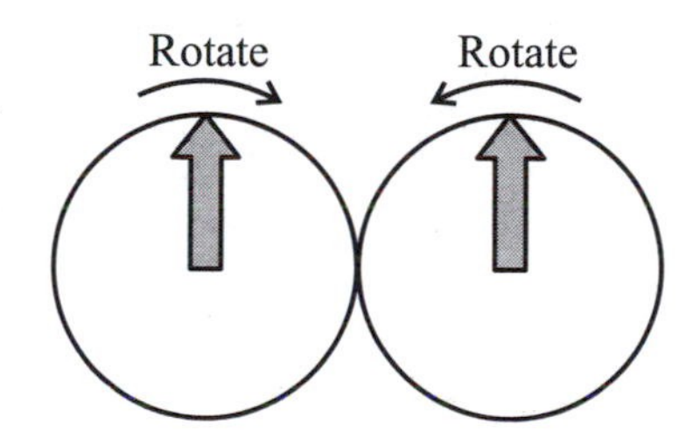

Cut out two large circles of equal size from thick cardboard and mark on each an arrow emanating from the center. Place the circles side by side with arrows pointing up and rotate each circle in the directions indicated. It takes a full rotation from each wheel before the two arrows are again parallel and pointing upwards. Notice each wheel "rolls" along the entire circumference of the other in this process.

Now hold one wheel fixed and roll the other wheel half way along its circumference. What happens?

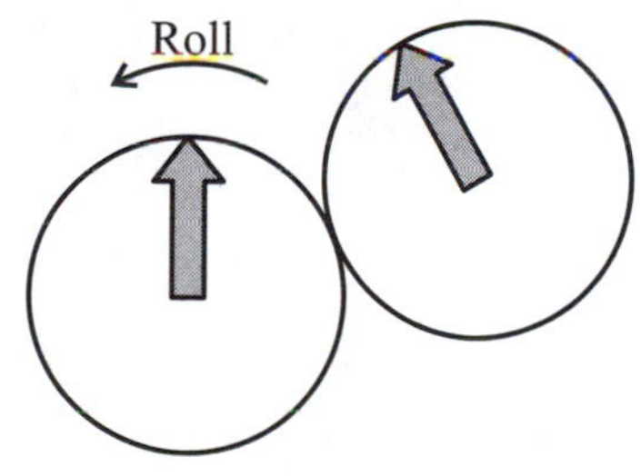

22.2 Freaky Wheels II

This puzzler was known to Aristotle more than 2000 years ago.

Cut two circles of different sizes from thick cardboard and glue the smaller wheel onto the larger so that their centers align. Mark a common radius on both wheels. Imagine these two wheels rolling along a double track as shown.

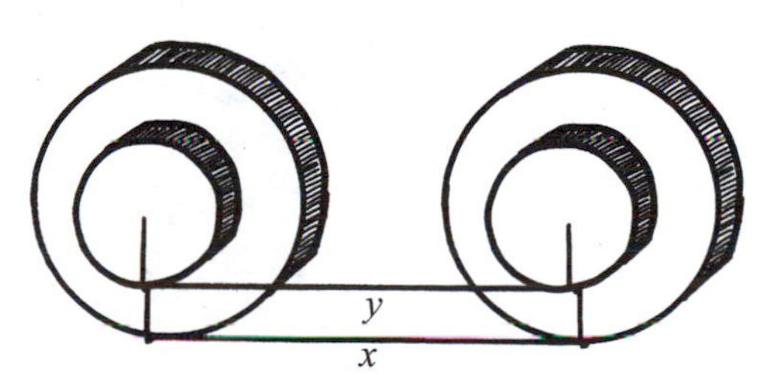

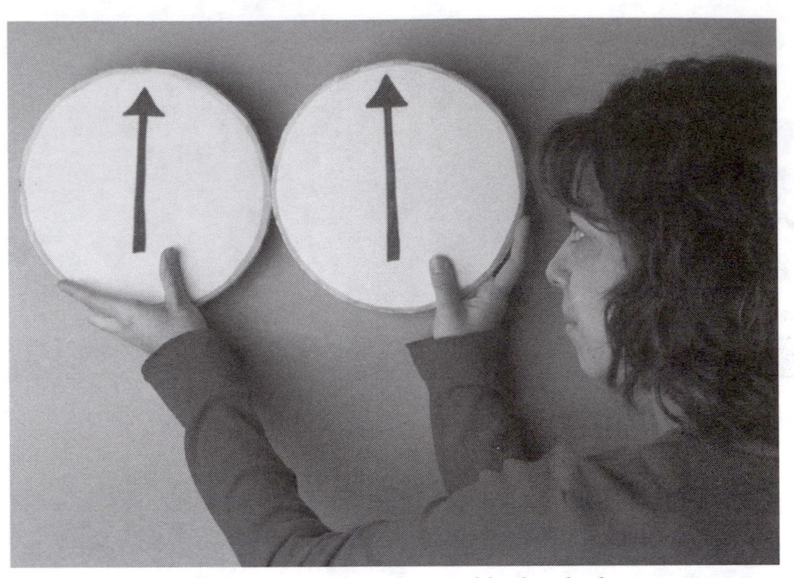

Kelly Ogden holds up a pair of freaky wheels.

Now consider one revolution of the entire system. Both wheels move along the same distance of track, so, in the diagram, $x = y$. But x equals the circumference of the big wheel and y the circumference of the little wheel. Is it true that these wheels have the same perimeter? Try the experiment to see that it must be the case!

John Dockstader experiments with Aristotle's paradox.

22.3 Breaking a Necklace

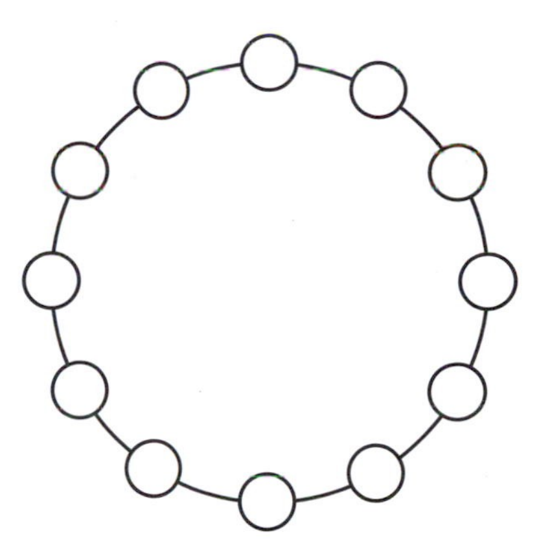

Here are 12 dots arranged in a circle. Color any six of them black. Two pirates have acquired a necklace containing six black pearls and six white pearls coincidentally arranged in the order you just colored! They would like to cut the necklace into as few pieces as possible so that, after divvying up the pieces, each pirate receives exactly three black and three white pearls. What is the minimal number of cuts they could make? Where should these cuts be placed?

22.4 Congruent Halves

This L shape can be cut into two congruent halves. (Mirror images are considered congruent.)

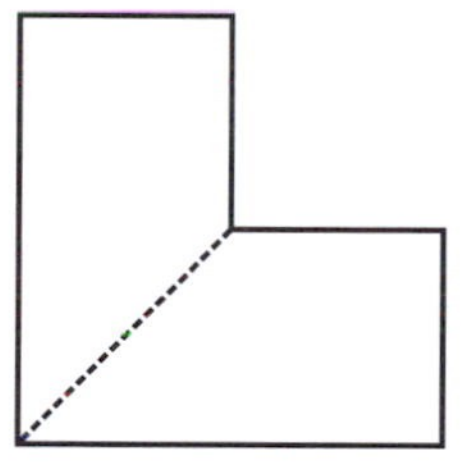

Which of these shapes can also be so subdivided?

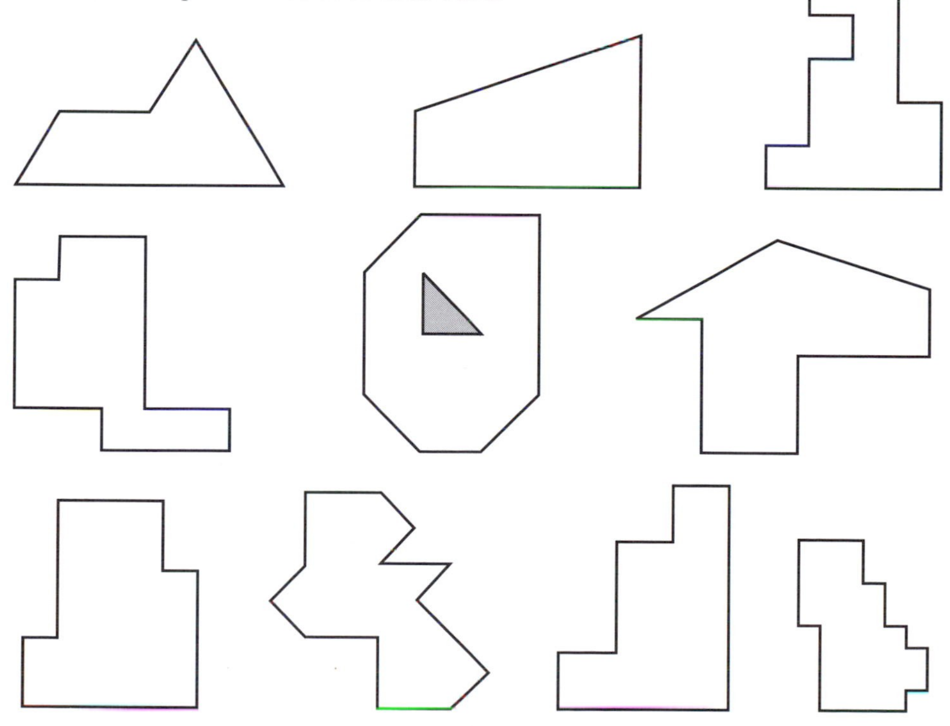

23

Playing with Playing Cards

23.1 A Pastiche of Card Surprises

Surprise 1. Shuffle a deck of cards and divide them into two piles of 26 cards each. Count the number of black cards in the first pile and the number of red cards in the second. What do you notice?

Surprise 2. Shuffle a deck of cards and divide them into two piles, one of 32 cards, the other of 20 cards. Count the number of black cards in the first pile and the number of red cards in the second. Subtract the smaller number from the larger. What do you get?

Surprise 3. Split a deck of cards into two piles according to color. Take ten cards from the red pile and shuffle them into the black pile. Without peeking, select ten cards from the (mostly) black pile and place them into the red pile. Count the number of "foreign" cards in each pile. What do you notice?

Surprise 4. Shuffle a deck of cards, take note of the top card, and place the deck face down on the table. Cut the deck, flip the top pile over, and place it back on the deck. Cut the deck again, deeper this time, and again flip the top pile over and place it back on the deck. Now remove all the cards that face up. What's the next card?

Surprise 5. Arrange a deck of cards so that all cards of the same suit appear in order from Ace, King, Queen, down to two. Each suit contains 13 cards. Cut the deck 13 times. Deal out, face down, a row of 13 cards from left to right. On top of them deal another row of 13 cards from left to right. Repeat this two more times until all the cards are dealt. Flip over each pile of four. What do you notice?

Surprise 6. Shuffle a deck of cards and note the bottom card. Call this the "magic card." Deal 12 cards face down and have someone turn over any four cards. Place the remaining eight cards on the bottom of the deck. Assigning the value 1 to an Ace, 10 to a Jack, Queen, or King, and the face value to a number card, sum the values of the four selected cards. Call this the "magic number."

On top of each selected card deal, in turn, the number of cards required to increase the face value of that card to 10. For example, on top of a 2 you would place eight cards, on top of a 7 three cards, and on top of a Jack zero cards. Collect all four piles of cards and place them on the bottom of the deck.

Now deal out the "magic number" of cards from the top of the deck. What's the final card dealt?

Surprise 7. Lay out 21 cards face up in a grid of seven rows and three columns. Have a friend mentally select one card and indicate to you the column in which that card lies. Pick up the cards one column at a time, carefully preserving the order within the columns, and making sure to collect the indicated column second.

Lay out the 21 cards again, row by row, to obtain seven rows of three. Have your friend again indicate to you the column in which the selected card lies. Pick up the cards as above and repeat this process one more time.

Pick up the cards again as above. Deal ten cards from the top of the pile and hand the 11th card to your friend!

23.2 Curious Piles

Shuffle a deck of cards and divide them into two equal piles of 26 cards. Clearly it is always possible to select a red card from one pile and a black card from the other. But suppose these cards were dealt instead into four piles of 13 cards. Would it be possible to select a spade from one pile, a club from another, a heart from a third, and a diamond from a fourth? (Try it!)
Is it always possible to accomplish this feat no matter how the cards are distributed?

23.3 On Perfect Shuffling

A *riffle shuffle*, also called a *Faro shuffle* or a *perfect shuffle*, begins by splitting a deck of cards precisely in half and then alternately interleaving the two halves back into a single pile. Thus, if one half of the deck were all the black cards, and the other half all the red, a riffle shuffle will result in a pile of alternating red and black. Dr. Brent Morris, a mathematician at the National Security Agency, is a master at perfect shuffling. He explains how you can perform it yourself on a full deck of cards in his wonderful book [Morr]. For ease, let's just work with eight cards numbered 1 to 8. Splitting the deck in half yields two piles, 1 2 3 4 and 5 6 7 8.

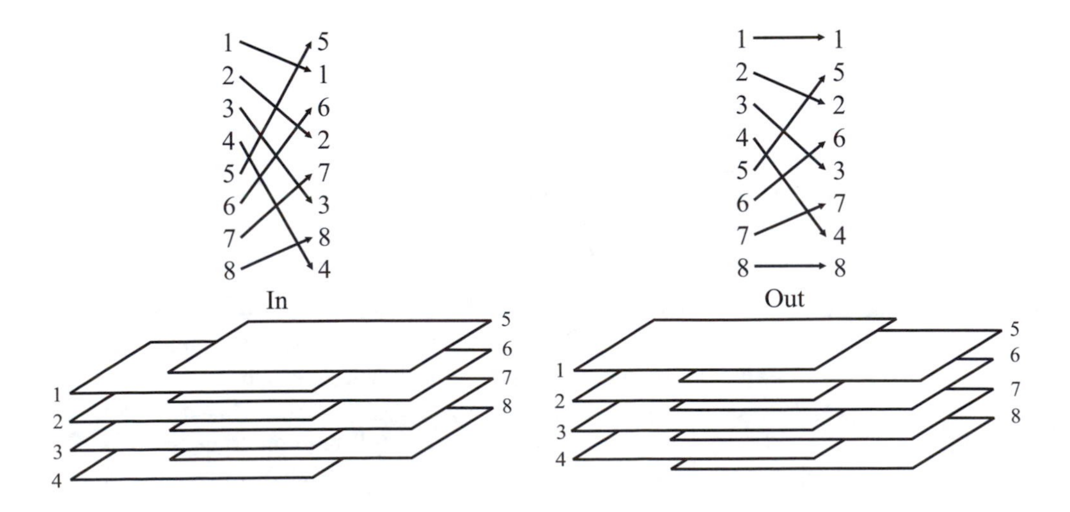

There are two types of perfect shuffles one can perform: an *in-shuffle*, which results in the ordering 5 1 6 2 7 3 8 4, and an *out-shuffle*, yielding the arrangement 1 5 2 6 3 7 4 8. Lay the cards in two rows of four and pick them up again by column to demonstrate these procedures.

What happens if you perform a perfect out-shuffle three times on a deck of eight cards? Using a combination of in- and out-shuffles, is it possible to move the top card of a deck of eight cards to an arbitrary position of the deck?

24

Map Mechanics

24.1 Cartographer's Wisdom

A map consists of *regions* bounded by *edges* that meet to form *vertices*. When painting a map, cartographers follow the convention that no two distinct regions sharing a common edge are assigned the same color (though two regions sharing a common vertex may be).

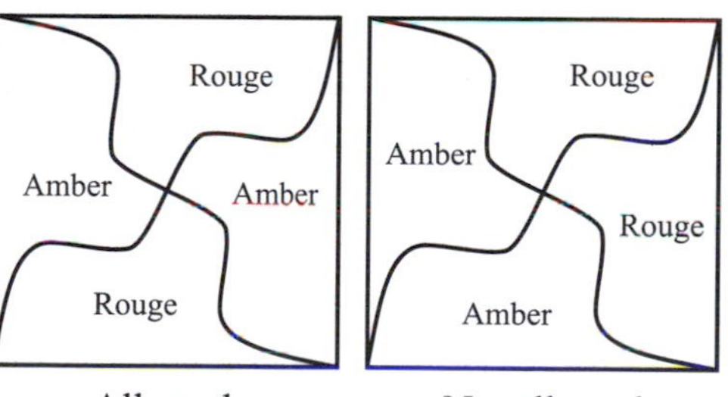

Allowed Not allowed

Cartographers have known for centuries (though see [May]) that just four colors are sufficient for coloring any map drawn on a plane. Some maps may be colored with less, but all maps can certainly be done with four. Try coloring this planar map with just four colors.

24.2 Simple Maps

This map is composed of regions arising from straight lines drawn across the entire page. It can be colored with fewer than four colors. How many fewer?

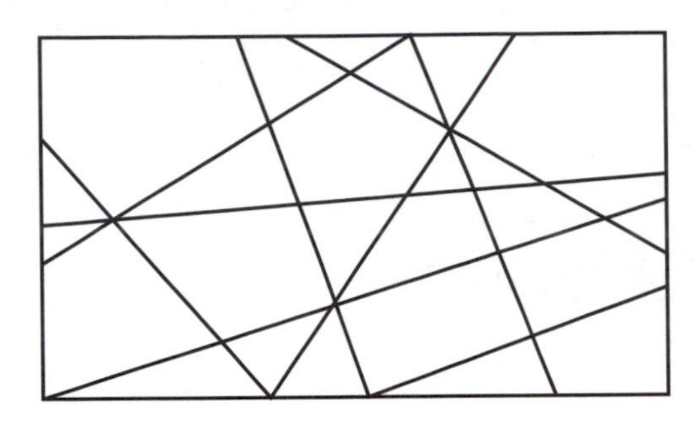

Taking It Further. How would your coloring scheme change if an extra line were added to the diagram?

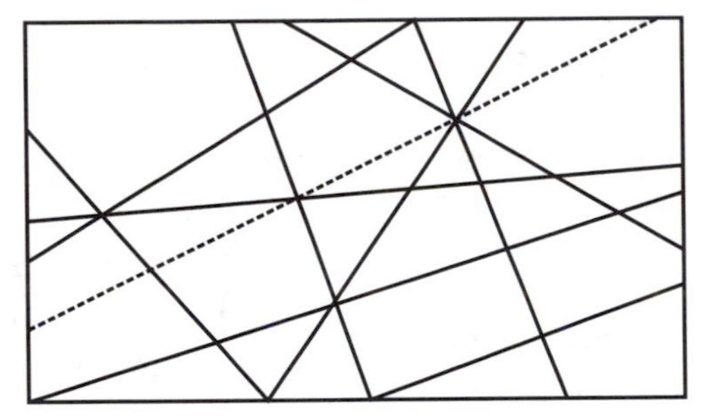

24.3 Toroidal Maps

Four colors are sufficient to color any map drawn on the plane. This implies that four colors are sufficient for coloring any map on a sphere as well: Simply puncture the sphere at the interior of any region, flatten out the map, paint it as though it were a planar design, and then reform the sphere.

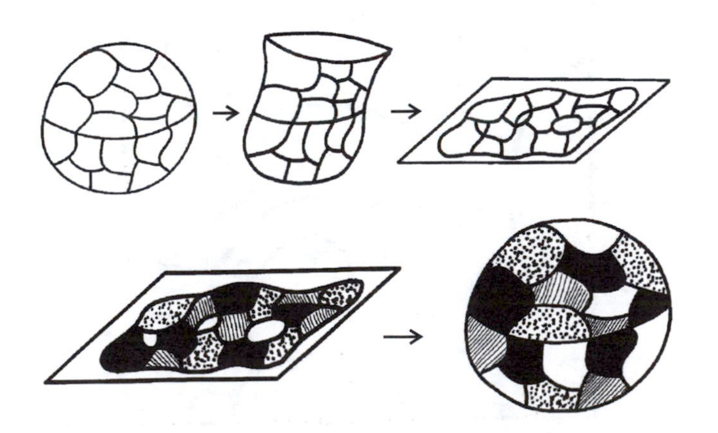

Four colors are also sufficient to paint any map on a cylinder. Can you see why?

But for maps drawn on a *torus* the situation is different. Can you design a map on a torus that requires a minimum of *five* colors to paint?

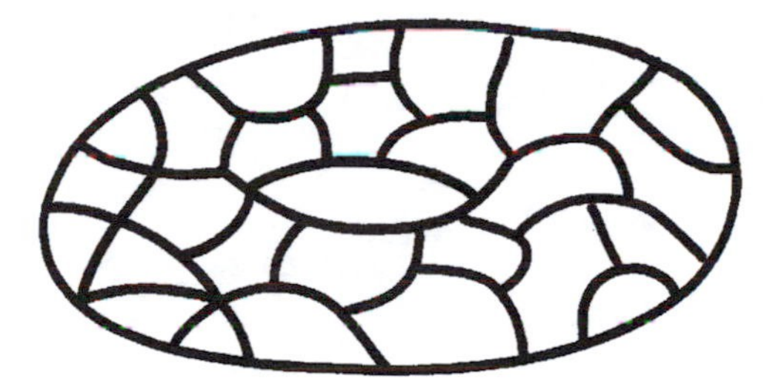

Try drawing and painting maps on real donuts or bagels!

25

Weird Lotteries

25.1 Winning Cake

A large group of people play an unusual lottery in the hopes of winning a cake. All write on a piece of paper their name and a positive integer greater than or equal to 1. All the entries are then collected and sorted through. If two or more people enter the same number, they are disqualified from the lottery. Only the unique numbers submitted are considered. The highest unique number wins, and the prize is that *fraction* of the cake! Thus if someone wins with the number 20 they win one twentieth of the cake and no one else will receive any cake. If you were to play this game, what strategy would you employ?

Comment. Try playing this game multiple times to experiment with alternative strategies. Cupcakes make good prizes.

25.2 Unexpected Winner

Some students write their names on individual ballots and place them into a hat. The professor selects one ballot at random to determine the winner of a fabulous chocolate cake. However, the professor suddenly makes this surprising announcement: "I am going to wait two minutes before announcing the name of the winner. No-one, except me of course, knows who has won the fabulous cake. You have no way of guessing who the winner could be, and it will remain that way for the next two minutes. The name of the winner will be a complete surprise to you all. John has won the cake." The professor is then silent for two minutes. Has John won the cake?

25.3 Winning Tootsie Rolls®

Everyone in a room is to write on a card the word "Cooperate" or "Defect." Tootsie Rolls® will be distributed according to the outcome of the following scheme. If everyone "cooperates" each person will receive ten Tootsie Rolls®. If a mixture of people cooperate and defect, or everyone defects, then the cooperators will each receive five tootsie rolls and the defectors none. But if there is just a single defector, he or she will receive 60 Tootsie Rolls® and the others none.

	C	D
All Cooperate	10	–
Two or more Defect	5	0
One Defects	0	60

Given the rules of this game, how would you vote?

25.4 Buying Tootsie Rolls

Clarence and Denise each receive a large pile of pennies with which to "buy" candy. The price, however, varies from turn to turn according to the roll of a die.

Die face	Candies per penny	Cost of six candies
1	1	6¢
2	2	3¢
3	2	3¢
4	3	2¢
5	3	2¢
6	6	1¢

At every roll of the die, Clarence will buy one penny's worth of candy, Denise will buy six candies no matter the cost. Who in the long run gets the better deal? Try this experiment several times and see if you can detect who gets the most candies for the money.

26

Flipped Out

26.1 A Real Cliff-Hanger

Dorothy stands on the edge of a cliff; an infinite expanse of land is behind her. Taking one step forward would send her to her doom, whereas one step back would be a step toward safety. All of Dorothy's steps are precisely one foot long. Dorothy has gamely agreed to let her fate be determined by the flip of a coin. She will take one step forward if the result of a toss is heads, one step back if it is tails. If she survives the first toss, she is willing to do it again, and again, stepping forward and back one foot according to the toss of the coin. After an infinite number of tosses she hopes to be wandering off into the infinite expanse behind her. What are Dorothy's chances of survival?

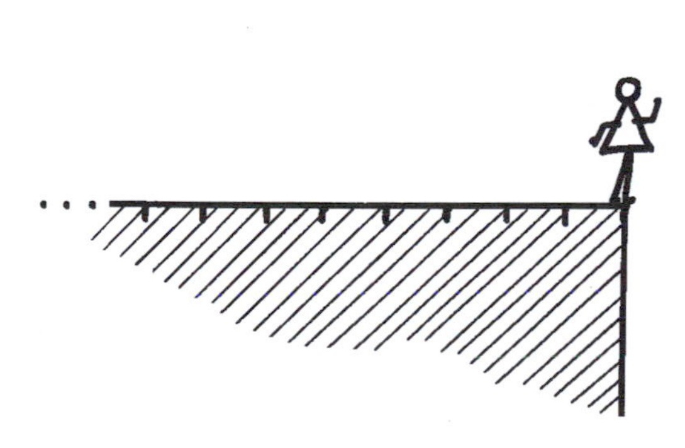

Angela Dellano tests Dorothy's fate with several repeated experiments.

26.2 Too Big a Difference

Toss a coin ten times. Then do it again, and again, many times. What is the average positive difference between the number of heads and the number of tails that appear? Is it zero?

26.3 A Surprise

Delicately balance 20 American pennies on edge on the surface of a table. Then bang the table so they all fall over. What do you notice?

Now simultaneously spin 20 American pennies and let them come to rest. What do you notice?

27

Parts That Do Not Add Up to Their Whole

27.1 A Fibonacci Mismatch

Who said area is always preserved? Take an 8×8 square inch piece of paper and subdivide it as shown. Now rearrange the pieces to form a 5×13 rectangle as shown. (Try it!) This transforms 64 square inches of paper into 65 square inches. What's going on?

Taking It Further. The Fibonacci sequence $1, 1, 2, 3, 5, 8, 13, 21, 34, 55, \ldots$ is defined recursively by

$$F_1 = 1$$
$$F_2 = 1$$
$$F_n = F_{n-1} + F_{n-2} \quad \text{if } n \geq 3.$$

Choose any Fibonacci number F_n equal to or larger than eight, and subdivide an $F_n \times F_n$ square as shown. Rearrange the pieces to produce an $F_{n-1} \times F_{n+1}$ rectangle. Have you again lost track of a square inch of paper?

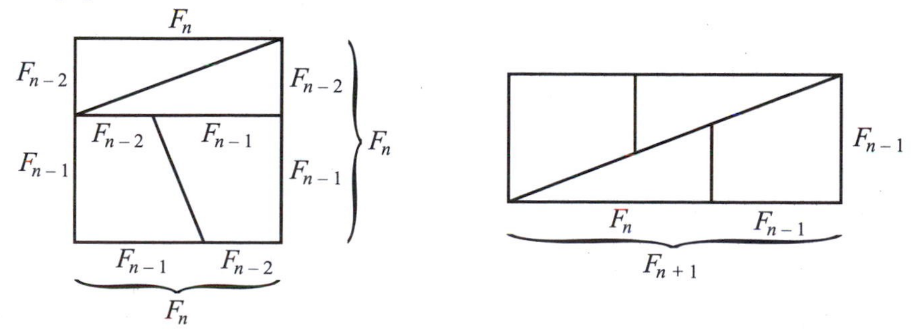

27.2 Cake Please

Two brothers, Albert and Hubert, plan to share a last piece of cake. They could perform the familiar "you cut – I choose" division scheme, where one brother cuts the slice into what he believes to be two equal parts and the other then chooses a piece. The first is then guaranteed, in his estimation, precisely 50% of the cake, the other, in his measure, 50% and perhaps more if he has a different estimation of half. It seems, however, the second person has an advantage in this scheme. Is there a cake slicing scheme that will guarantee *both* brothers more than 50% of the cake in their own estimations?

27.3 Sharing Indivisible Goods

Bjorn and Elaina each have a large supply of Tootsie Rolls® but only one chocolate bar between them. For some reason they will not break the bar in two, but both are willing to trade their share of the bar for Tootsie Rolls®. Bjorn thinks the bar is worth 18 Tootsie Rolls®, Elaina 14. Who should keep the bar? How many Tootsie Rolls® should the other person receive as compensation? Devise a scheme so that each person gets more than his or her estimation of worth!

28

Making the Sacrifice

28.1 The Josephus Flavius Story

Let's begin on a positive note.

A group of people sit in a circle and begin to count off every third person. Whoever is selected third is called "out" and leaves the game. Counting continues until the last participant is declared the winner of the game. Given the size of a group and where the count begins, is it possible to predict beforehand who will be the winner? Try playing the game several times with different sized groups and see if you can detect any pattern to the location of the winner.

Taking It Further. What if every second person is counted? Every ninth person?

28.2 Soldiers in the Desert

A horizontal line is drawn on an arbitrarily large grid of squares. Behind this line stands an army of pennies, one penny per cell. The aim is to move a single penny into the desert by performing a series of checker jump moves within the troops. One penny can jump over any other penny in either a vertical or a horizontal direction to a vacant square. The penny jumped over is removed or "sacrificed". It is easy to design an army that could move a penny one line or two lines into the desert. Is it possible to move a penny three lines into the desert? How about four lines into the desert?

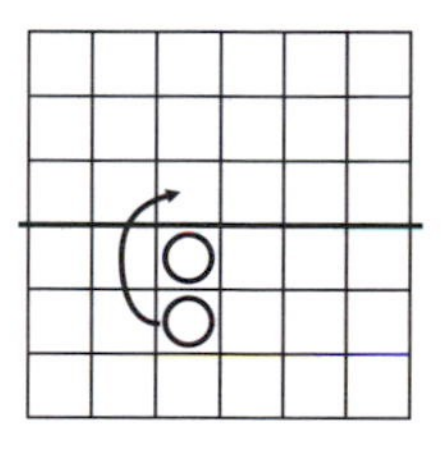

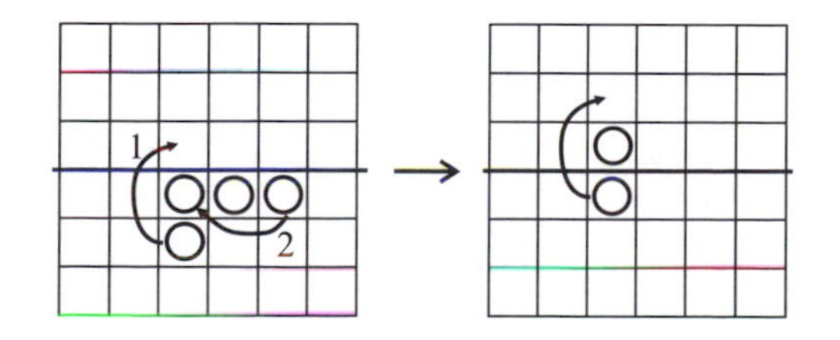

28.3 Democratic Pirates

Ten pirates ranging in rank from captain (whom we shall label pirate number 1) to cabin boy (pirate number 10) have come across ten gold coins. Being a democratic crew they decide upon the following process to distribute the coins. The cabin boy will first nominate a distribution pattern (two coins to pirates 1, 3, 4, 8, and himself, for example) and the pirates will take a vote. If 50% or more of the pirates agree with this distribution they will go with it, otherwise the cabin boy will be thrown overboard and the ninth ranking pirate will propose a new scheme and invoke a new vote. They will do this, up the rank, until they finally settle upon a favorable vote.

What distribution scheme should the cabin boy suggest? Assume the pirates are rational thinkers and none will vote against a proposed scheme if, as a consequence, he will be thrown overboard, or end up with fewer coins. If there is no personal ill effect, a pirate will otherwise be glad to see a fellow shipmate thrown overboard.

Comment. This game can be acted out using a supply of ten treats. Rather than ten pirates, consider first a game with just two or three pirates.

29

Problems in Parity

29.1 Magic Triangles

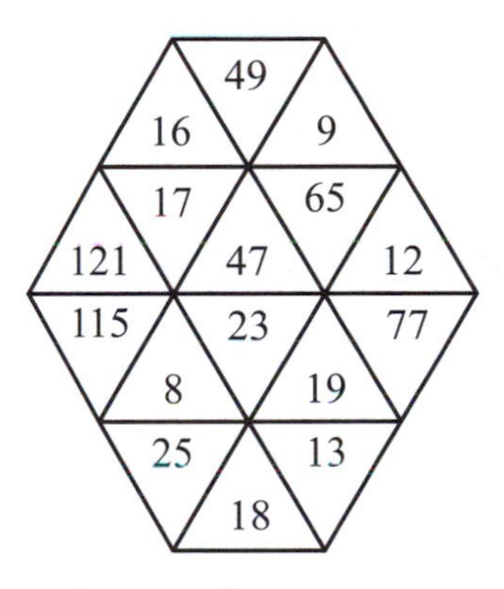

Consider the grid of triangles shown. Two triangles are said to be *neighbors* if they share a common edge. Thus 17 and 47 are neighbors, but 17 and 9 are not. A *path* in the grid is any sequence of neighboring cells. A path might loop back on itself or even step back and forth repeatedly between a select few cells. For example, 17-47-65-9-65-47-23 is a valid path of **six** steps (even though seven numbers are mentioned).

Place your finger on any even number in the grid and perform the following sequence of instructions.

1. Move your finger 11 steps along any meandering path of your choosing. Keep your finger in place at the end of your chosen path. You will start step 2 from here. (Despite the free will you possess I know you have not landed on a cell that is a multiple of 3!)

2. Avoiding all multiples of 3, move your finger seven steps along any path of your choosing. (I know you have not landed on a multiple of 5! I could make similar predictions as you continue.)

3. Avoiding all multiples of 3 and 5, move your finger six steps along any path.

4. Avoiding all multiples of 3, 5, and 7, move your finger five steps along any path.

5. Avoiding all multiples of 3, 5, 7, and 11, move your finger five steps along any path.

6. Avoiding all multiples of 3, 5, 7, 11, and 13, move your finger seven steps along any path.

7. Avoiding all multiples of 2, 3, 5, 7, 11, and 13, move your finger three steps along any path.

8. Avoiding all multiples of 2, 3, 5, 7, 11, 13, and 17, move one place over.

How do I know that you will never land on an inappropriate cell? How do I know that your finger is currently on cell number 23?

Comment. Make copies of this grid and perform the activity with a group. Everyone in the group will land on precisely the same cell in the end to their great surprise.

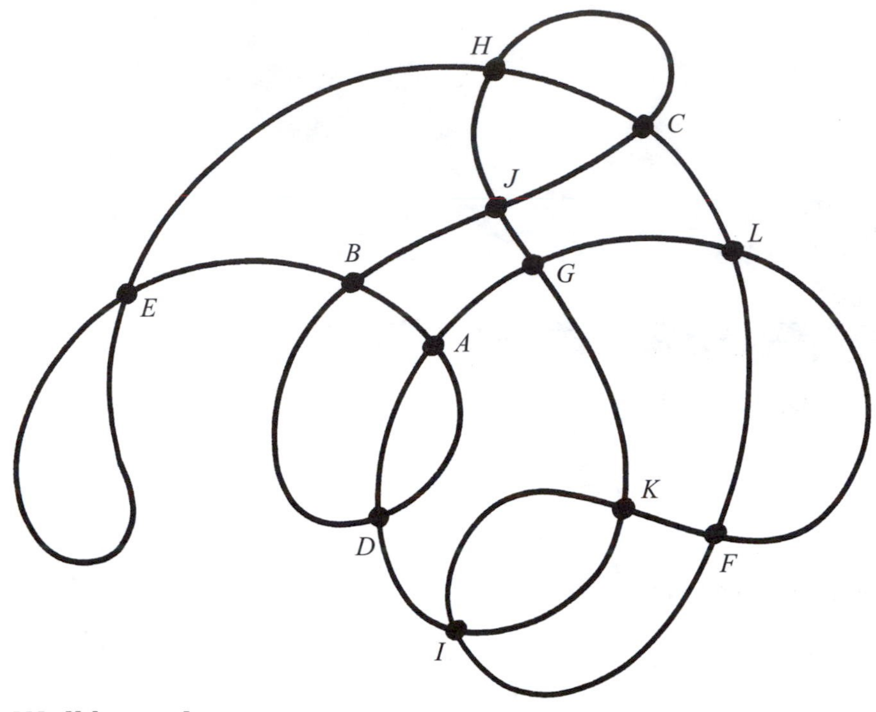

29.2 Walking a Loop

Here is a closed loop that crosses over itself several times. The curve has the property that it passes through each of its crossing points only twice. (One can draw curves that pass through the same point many more times, but we ignore curves of that type here.) I have labelled the crossing points randomly with the letters A through L. These letters can be used to record a journey that traverses the entire loop once:

A-D-B-J-C-H-J-G-K-I-D-A-G-
L-F-K-I-F-L-C-H-E-E-B-A

Have a friend secretly draw a curve of this type on a piece of paper and label the crossing points with letters in some arbitrary fashion. Now have the person trace the curve with her finger and read out to you the names of the crossing points she encounters, but with one deliberate error: she is to transpose the names of two adjacent crossing points. (For example, in the curve above, when I am at crossing point G and about to move to L I could read out L G rather than G L.) Your friend should never reveal to you the picture, the labelling scheme, or the place of the error.

Relying only on the sequence of letters said out loud, how could you swiftly determine which two crossing points were switched? Once you have mastered this trick, use it to impress your friends!

29.3 Catch Me If You Can

Here is an international island-hopping game. Craig starts in Perth, and Joy in Papeete. Each takes turns hopping from one location to the next along an established air route as in the dia-

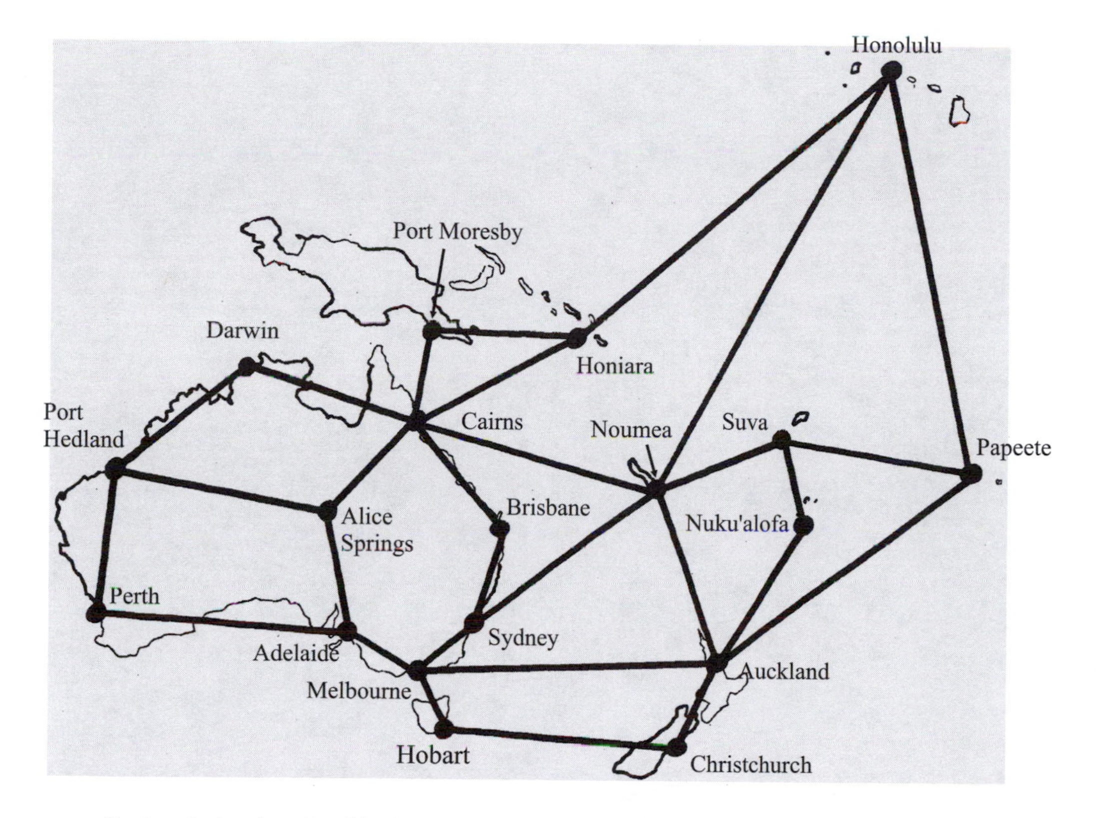

gram. Joy's mission is to land in the same location as Craig. Craig's mission is to avoid this. Joy goes first and has 12 turns in which to win this game. If she fails, Craig will be declared winner. What is Joy's best strategy in order to win this game?

29.4 A Game of Solitaire

This game of solitaire is played with a line of pennies placed heads or tails up in an arbitrary fashion. A move consists of removing any coin that is heads up and flipping over the coin in the space to its immediate left (if one is there to be flipped) and the coin in the space to its immediate right (if present). The goal is to remove all the pennies. What is the best strategy for winning this game of solitaire? When should you not even bother playing the game?

30

Chessboard Maneuvers

30.1 Grid Walking

Greta, Peter, Lashana, and Lopsided Charlie wander within a large grid of square paving stones, each following certain rules of motion.

Greta moves only in horizontal and vertical directions, one square over at a time. Can anything be said about the total number of steps she must take to return to her initial square and thus form a loop of steps?

Peter moves about the grid like a knight on a chessboard. Each step takes him two squares over in one direction and one square in an orthogonal direction. Can anything be said about the total number of steps he must take to complete a loop?

Lashana takes only single diagonal steps as she wanders about the grid. Can anything be said about the total number of steps she must take to complete a loop?

Lopsided Charlie follows slightly more complicated rules of motion. Any step he takes north or east moves him *two* squares, but any step south or west moves him only *one* square. Can anything be said about the total number of steps Charlie must take to complete a loop?

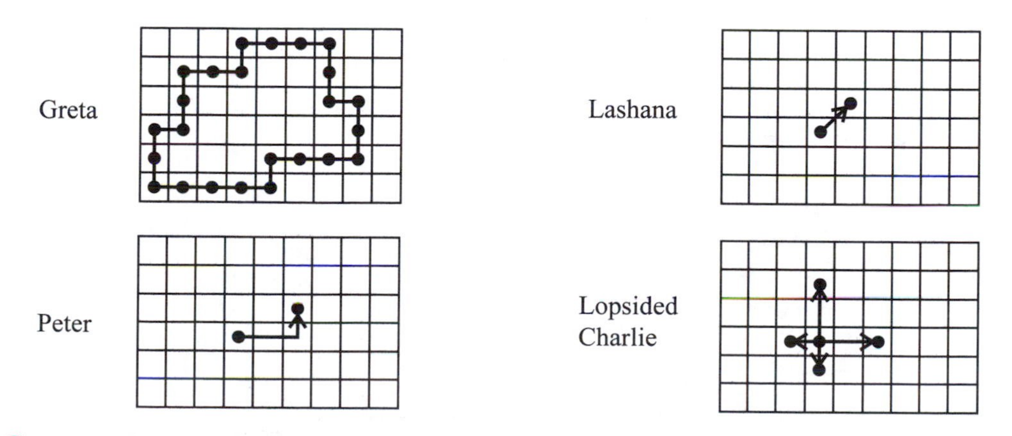

Comment. Use graph paper to experiment with any of the above motions.

30.2 Kingly Maneuvers

A king moves across a chessboard one square at a time in any one of eight directions: left, right, up, down, or diagonally. In this problem we want the king to move from the top row of an 8×8 chessboard to the bottom row solely on black squares, or from the leftmost column to the rightmost solely on white squares. Both are possible with the standard coloring scheme of a chessboard as shown.

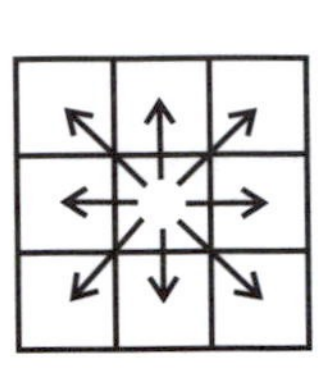

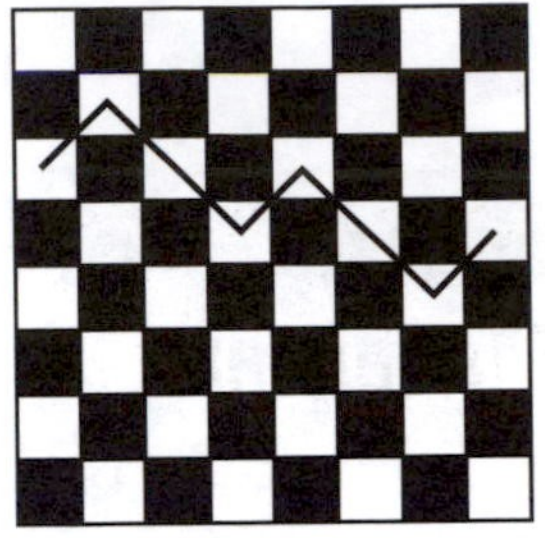

Make an 8×8 grid of squares and randomly color the cells black and white. (You need not have the same number of cells for each color). In your arbitrary coloring scheme, is there still a path of black squares from the top row to the bottom for a king to travel? If not, have you unwittingly created instead a path of white cells from left to right? Devise an 8×8 coloring scheme that creates no paths of either type.

Comment. Turn this puzzle into a two-person game. Have two players take turns coloring the cells of an 8×8 grid, the first player coloring the cells black, the second white. The first player wins by creating a path of black cells connecting the top row to the bottom, before the second player creates a path of white cells linking the leftmost column to the rightmost (otherwise the second player wins.) Must there be a winner for every game?

30.3. Mutual Non-Attack: Rooks

A rook attacks other pieces on a chessboard by sliding along vertical or horizontal lines of squares and taking the place of any opponent it encounters. What is the maximal number of rooks that can be placed on a 4×4 chessboard so that no rook is in position to attack any other rook?

What is the maximal number of rooks that can be placed on an $n \times n$ board in a configuration of mutual non-attack? What about $p \times q$ rectangular boards with $q < p$? How many different ways are there to place a maximal number of rooks on an $n \times n$ board in mutual non-attack? On a $p \times q$ board?

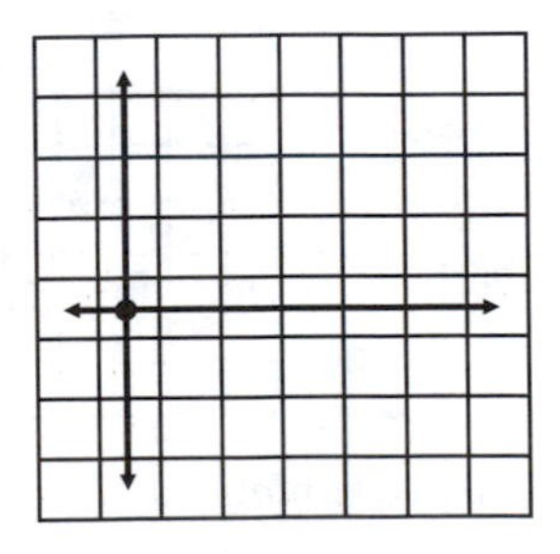

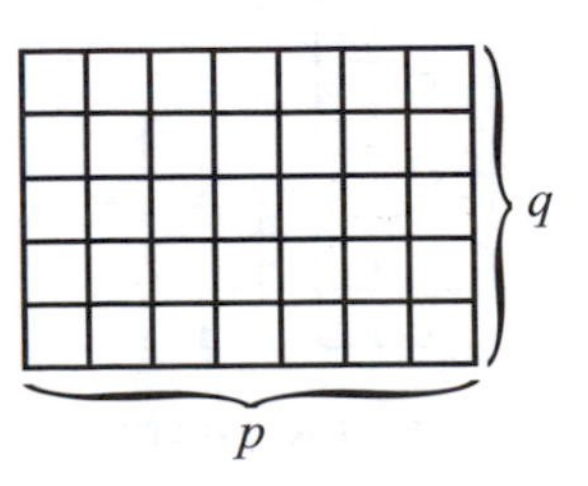

30.4 Mutual Non-Attack: Queens

A queen attacks other pieces on a chessboard by sliding along vertical, horizontal, or diagonal lines of squares and taking the place of any opponent encountered. What is the maximal number of queens that can be placed on a 4×4 chessboard so that no queen is in position to attack another queen? What about 5×5, 6×6, and 8×8 chessboards?

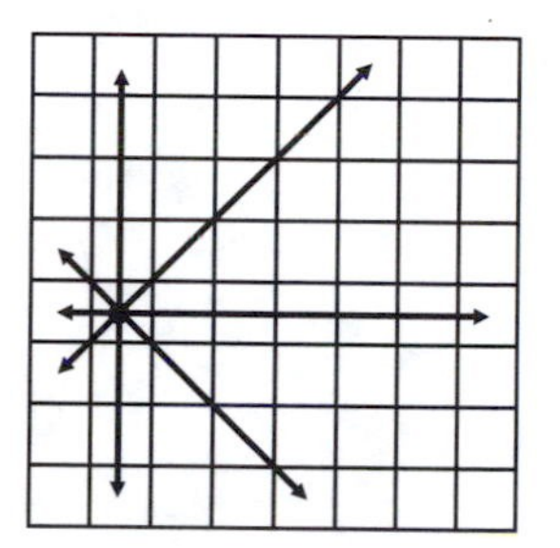

PART II

HINTS, SOME SOLUTIONS, AND FURTHER THOUGHTS

1 Distribution Dilemmas

1.1 A Shepherd and His Sheep

Many puzzles require the reader to think beyond the boundaries suggested (but not enforced!) by the problem. Connecting dots in a 3×3 array with contiguous straight line segments is a prime example of such a puzzle: *It is possible to connect nine dots in a* 3×3 *array with five contiguous line segments. It can also be done in four. Can you see how?*

These inheritance puzzles similarly require stepping beyond the boundaries implied. As a hint, what is the lowest common multiple of 2, 3, and 9? What is the sum $\frac{1}{2} + \frac{1}{3} + \frac{1}{9}$?

Comment. The three sons could, of course, convert the flock to mutton and divvy up the meat according to weight. But what should the sons do with the leftover mutton—divvy it up again according to the same proportions?

1.2 Iterated Sharing

At the end of each iteration do two things. First, count the number of people with the smallest amount of candy. Second, note the largest amount of candy any one person possesses. What do you notice?

2 Weird Shapes

2.1 Plucky Perimeters

Ignoring units, their perimeters equal their areas!

Taking it Further. Is there a rectangle with this property? An equilateral triangle?

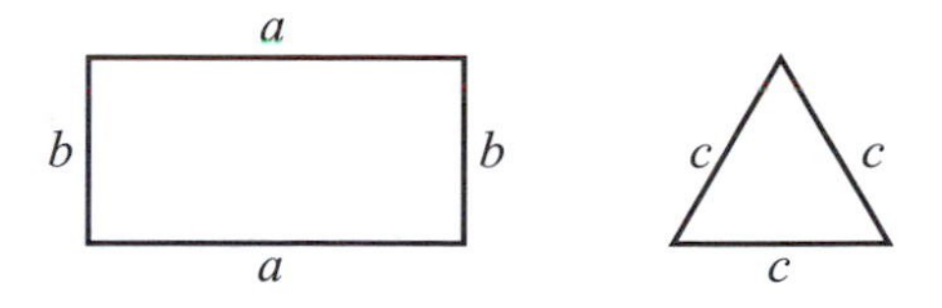

Only one other right triangle with integer side lengths has this property. What is it?

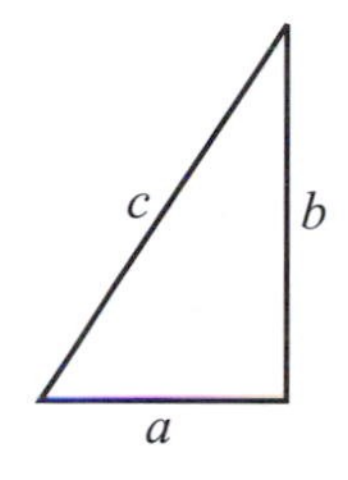

Is there a (non-circular) ellipse with this property?

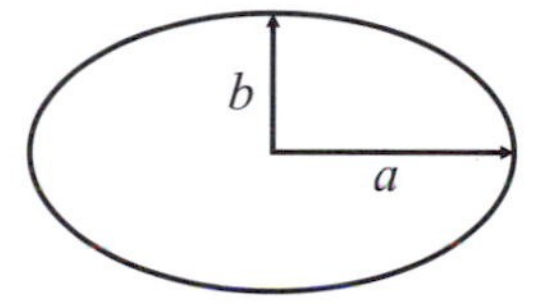

Taking it Even Further. Is there a rectangular box whose volume equals both its surface area and the total sum of its edge lengths?

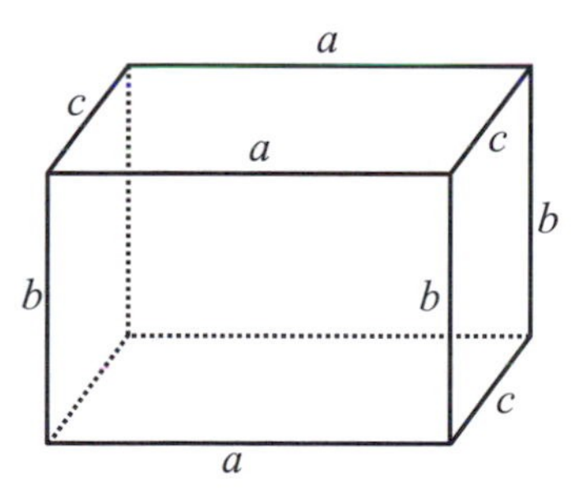

2.2 Weird Wheels

(*WARNING:* I used some sophisticated college mathematics!) The parametric equations of a circular wheel, unit radius, are:

$$(0 + \cos\theta, 0 + \sin\theta) \quad 0 \le \theta \le 360^\circ.$$

In these equations I have made the center of the circle, (0, 0), explicit. This certainly has the property that the distance between any two points 180° apart is always 2.

We can modify these equations by constantly changing the location of the center of the circle. As long as we ensure that the center returns to the same place after 180°, the distance between two points that are this angle apart will again always be 2. For example,

$$\left(\tfrac{1}{8}\sin(2\theta)+\cos\theta, \tfrac{1}{10}(2\theta)+\sin\theta\right) 0 \le \theta \le 360^\circ$$

does the trick. (In fact these were the equations I used to generate the shape given.)

How could you achieve the same effect *without* the use of a computer and sophisticated mathematics?

2.3 Square Pegs and Not-so-Round Holes

Given the hint and solution to Problem 2.2, modifying the equation of a circle so that its center "wobbles" could conceivably produce new shaped holes with the desired property. We would want the wobbling center to return to the location after each 90° rotation, given the rotational symmetry of a square. Is there a simple ruler and string construction that would do this for us? Can we produce these shapes with the aid of a computer?

3 Counting the Odds ... and Evens

3.1 A Coin Trick

When the coins are first tossed, Han quickly checks to see whether an even or odd number of heads is showing. He counts the number of times the word "flip" is uttered and again checks the evenness or oddness (that is, the *parity*) of the number of heads when he opens his eyes. This gives Han enough information to determine the state of the coin covered by John's hand.

3.2 Let's Shake Hands

Suppose N people are in the room. Let n_1 be the number of times the first person shakes hands, n_2 the number of times the second person shakes hands, and so on. What can you say about the sum

$$n_1 + n_2 + \cdots + n_N?$$

Can all the numbers n_i be odd?

3.3 Forty Five Cups

Initially the number of upright cups is odd. What do you notice about the number of upright cups after each move?

3.4 More Plastic Cups

A cup in the nth position is turned over once for every divisor d of n. Which numbers have an odd number of divisors?

4 Dicing, Slicing, and Avoiding the Bad Bits

4.1 Efficient Tofu Cutting

A $2 \times 2 \times 2$ cube of tofu, for example, cannot be diced into eight smaller cubes in fewer than three planar slices. Each corner cube has three

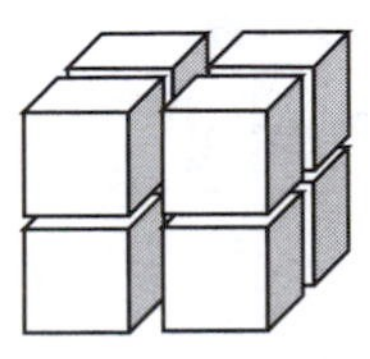

faces that must be cut, and no single planar slice will cut more than one of them.

4.2 Efficient Paper Slicing

It is helpful to take the problem down yet another dimension! With stacking allowed, what is the minimal number of cuts needed to divide a piece of string n units long into n unit segments?

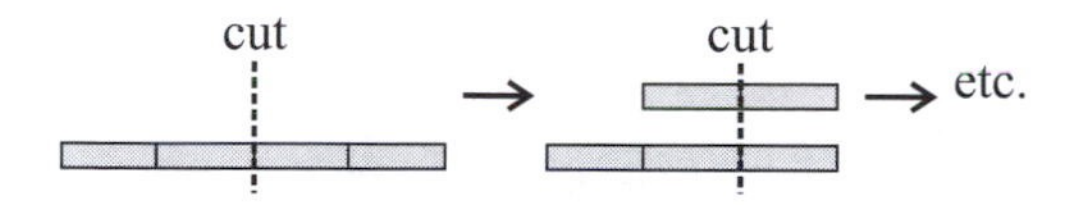

4.3 Bad Chocolate (Impossible!)

For the game with the 4×8 bar, I advise you to be the first player. In the 4×4 game, however, I strongly recommend your being the second player.

5 “Impossible Paper Tricks

5.1 A Big Hole

Begin by cutting a slit along a center line and then folding the card in half. Make some more cuts. Think about it.

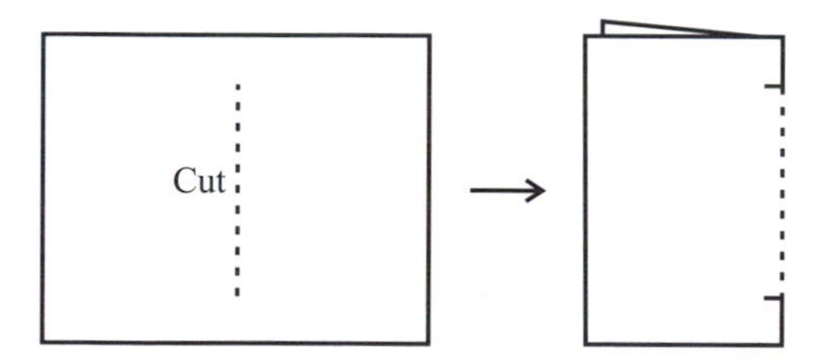

5.2 A Mysterious Flap

Begin by taking an ordinary sheet of paper and make three cuts to the center line as shown. Now what?

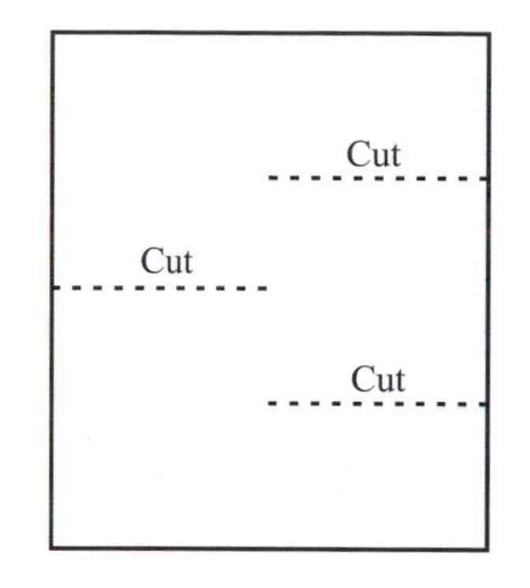

5.3 Bizarre Braids

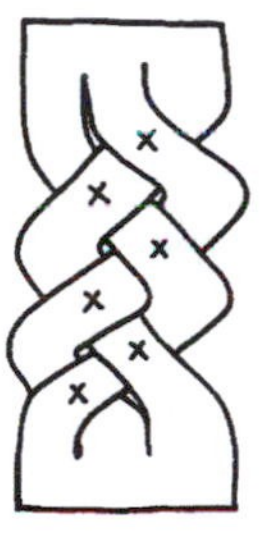

Notice that in the braid there are six places where two strands cross over. This is necessarily the case if the left strand is to return to the left position, the middle strand to the middle, and the right strand to the right. Hold the paper (or felt) in front of you with both hands and begin braiding from the top, ignoring whatever happens to the bottom half of the paper. Perform five crossovers and hold the fifth one firmly with one hand. *Do not let go!* With the other hand, do whatever it takes to untangle the bottom half. In the process a sixth crossover point will occur.

Taking it Further. Is it possible to make a four-strand braid with no free ends?

5.4 Linked Unlinked Rings

Yes it is! The diagram on the next page is flat. In three dimensions, however, you can place one ring flat in the xy-plane, the second flat in the yz-plane and the third flat in the xz-plane. This leads to a particularly pleasing arrangement. It is fun to construct this design with the aid of two

friends. One person forms a ring with the thumbs and index fingers from each hand. The other two people then do the same but along orthogonal planes about the first person's hands. (Try it!)

Comment. These three rings are known as the Borromean rings. They appeared on the coat of arms of the famous Italian Renaissance family of Borromeo.

Taking it Further. Is it possible to link *four* rings of paper so that on any single cut the whole configuration will separate into four pieces?

Aside. Ring snipping reminds me of a puzzle. You have a gold chain of six links and have agreed to give to your friend one or more or perhaps even all of the links. The roll of a die will decide the number. Which one link could you cut that would ensure your ability to pay your friend no matter the outcome on the die?

6 Tiling Challenges

6.1 Checkerboard Tiling I

Color the 7×7 array according to a standard checkerboard scheme of 25 black cells and 24 white cells. As each domino covers one cell of each color, any tilable configuration of 48 squares must be composed of an equal number of black and white squares. Excising a white cell thus leaves an **un**tilable diagram. What can be said about removing a black cell?

6.2 Checkerboard Tiling II

Again consider the standard checkerboard coloring scheme. Two cells of the same color have been excised leaving a configuration with an unequal number of black and white cells, rendering it untilable.

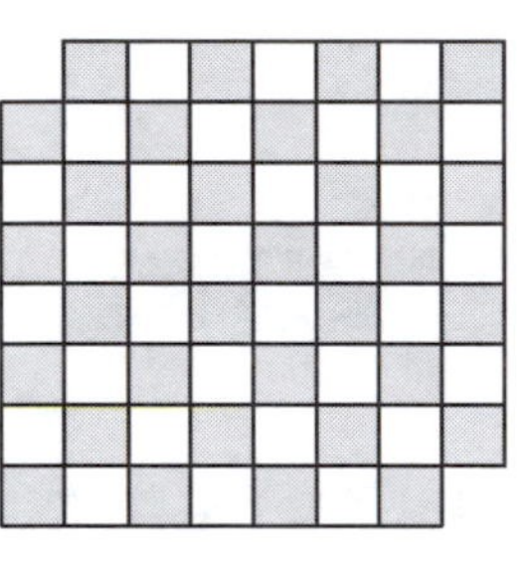

Taking it Further. Two arbitrary cells of opposite color are excised from an 8×8 checkerboard. Is the remaining configuration guaranteed to be tilable?

6.3 Checkerboard Tiling III

It is possible. The key lies in the placement of the single 1×1 tile. Consider these two coloring schemes of the 8×8 grid to help figure out just where that single tile should go.

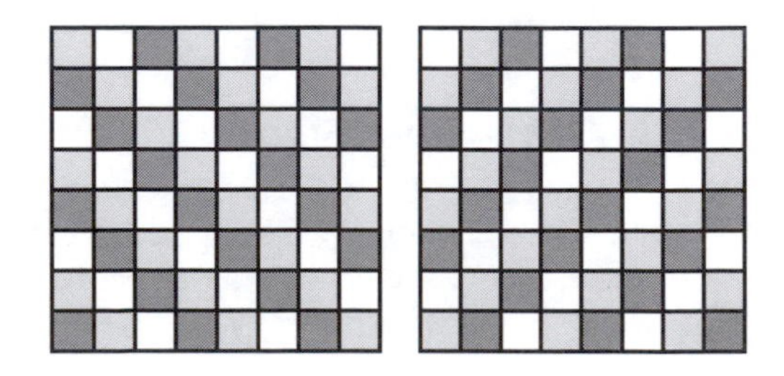

6.4 Checkerboard Tiling IV

Having trouble? Maybe such a tiling does not exist. How many dominoes would you need to ensure that each potential separating line is crossed by a tile?

7 Things That Won't Fall Down

7.1 Wildly Wobbly

Of course not. Any such figure would perpetually topple from one face to the next and never come to rest. Perpetual motion devices do not exist.

Taking It Further. A cube has six stable faces; it sits at rest no matter which face it is on. The figure on page 15 has five stable faces and one unstable face (assuming of course that the figure is made from uniformly dense material with no hidden weights or hollows).

Is it possible to design a six-faced polyhedron with precisely two unstable faces? How about one with three? Four? Does there exist such a figure with five unstable faces? Try carving models of six-faced polyhedra out of florist's foam.

Taking It Even Further. Experiment with the two-dimensional analog of this problem by studying the motion of polygonal wheels. Given how wheels roll, it is appropriate to assume our wheels are convex in shape. As perpetual motion devices do not exist, every wheel must possess at least one stable edge. Is it possible to design a polygonal wheel with *precisely one* stable edge? The wheel may have as many sides as you wish.

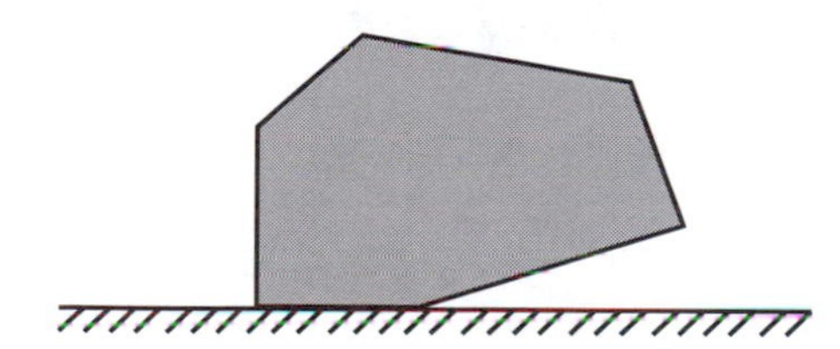

7.2 A Troubling Mobile

A single wire is balanced with a thread attached to its midpoint. The first challenge is to find where a two-wire system will balance if the other end of the thread is attached to a second wire.

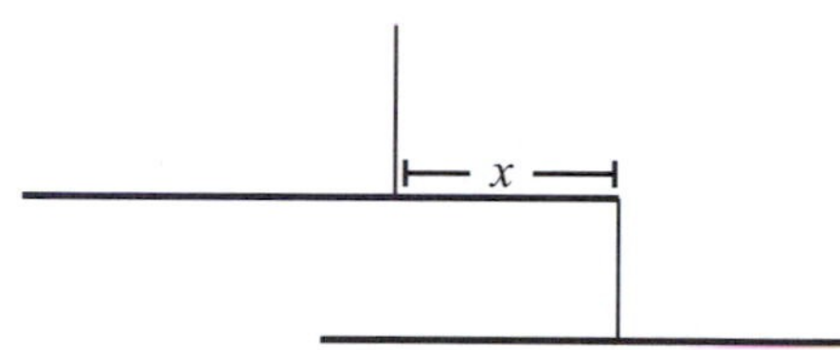

Archimedes' Law of the Lever says that two weights m_1 and m_2 at distances x_1 and x_2 from the fulcrum will balance precisely when $x_1 m_1 = x_2 m_2$. If each wire is one foot long and

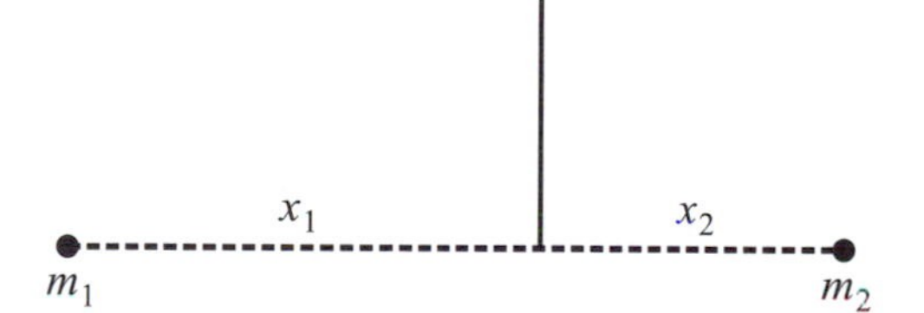

has mass m the two-wire system is equivalent to two balanced masses, one being the center of mass of the top wire at its midpoint, and the other the mass of the lower wire at the end. (See figure below.) Archimedes' Law thus dictates

$$\left(\frac{1}{2} - x\right)m = xm,$$

forcing $x = 1/4$.

At which point will a three-wire system balance?

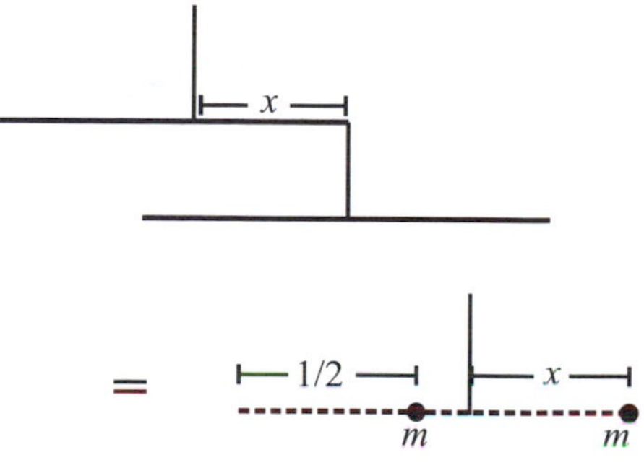

7.3 A Troubling Tower

This problem is an upside-down version of 7.2, and its solution is essentially the same.

8 Möbius Madness: Tortuous Twists on a Classic Theme

8.1 Möbius Basics

A Möbius band is formed by gluing together the ends of a long strip of paper in reverse orientation. Notice that the two segments marked A are glued together, as are the remaining two line segments labelled B. How many pieces will result, then, when this diagram is cut along the center line?

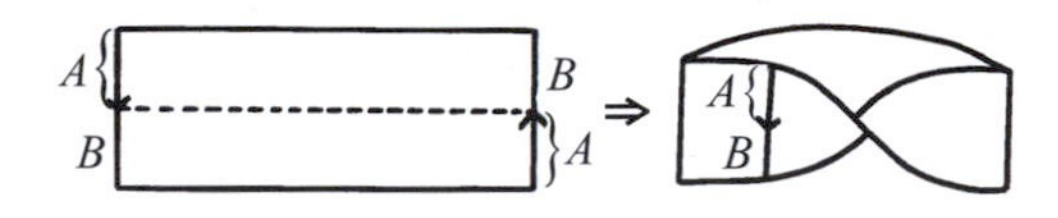

8.2 A Diabolical Möbius Construction

This construction is really a simple Möbius band with the addition of a single connecting strip. Under what conditions will this figure separate into two distinct pieces when cut?

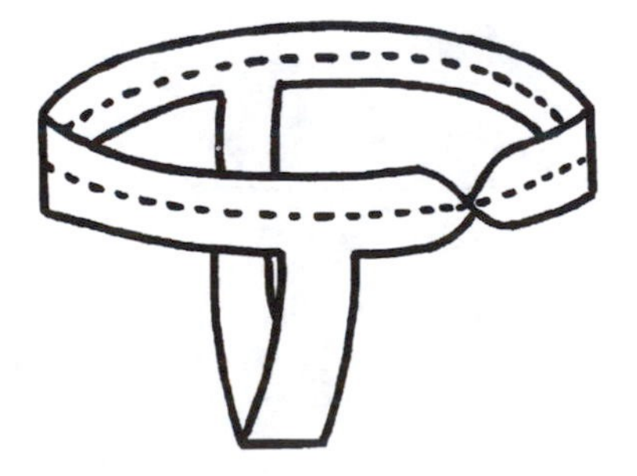

8.3 Another Diabolical Möbius Construction

Cut around the non-twisted band first.

9 The Infamous Bicycle Problem

9.1 Which Way Did the Bicycle Go?

Think of how a bicycle is constructed. The back wheel is fixed in its frame yet the front wheel can turn and even wobble. We thus deduce that the more stable track is the back wheel track and the other the front wheel track. Moreover, the back wheel of a bicycle is fixed in its frame so as to *always point towards the front wheel.* What does this imply about the structure of the two curves?

9.2 Pedal Power

The pedal is being pushed in a direction that would normally drive the bicycle forward. So the bicycle moves forward, right? Try it on a bicycle and see what happens!

9.3 Yo-Yo Quirk

Again, try it! See what happens.

10 Making Surfaces in 3- and 4-Dimensional Space

10.1 Making a Torus

First form a cylinder by gluing the top edge A to the bottom edge A. What must you then do

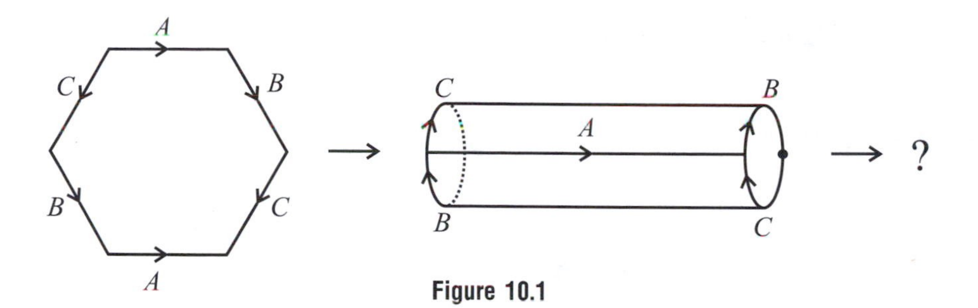

Figure 10.1

to this cylinder to complete the edge identifications? See Figure 10.1.

10.2 A Torus with a Serious Twist

This half twist changes the topology of the problem significantly. If you attempt this feat, you will notice that the two ends of the tube being created end up lying on opposite sides of the band. It is not clear what to do next to complete the edge identification.

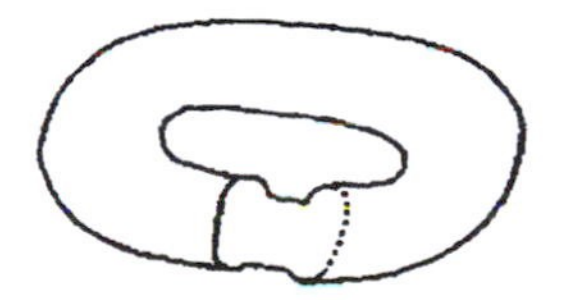

Here's another way to think about the problem. A *Möbius band* is obtained from a rectangular strip of paper by gluing the two ends together, here labelled A, with reverse orientation.

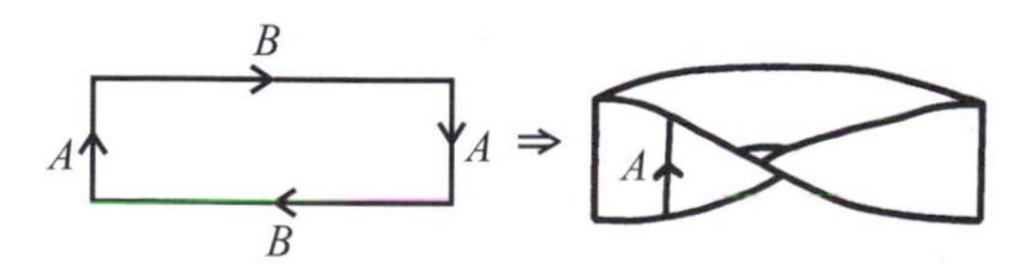

In this exercise we are asked to then glue together the edges marked B. Try instead gluing the B edges together first to form a cylinder, and then the edges marked A. Again, make sure the directions of arrows match. See Figure 10.2.

10.3 Capping Möbius

Consider the following question: Is it always possible to sew the boundary of a flexible disc onto a curve drawn in a plane? The answer is yes, but you may have to move into the third dimension to do it. Regarding a disc as com-

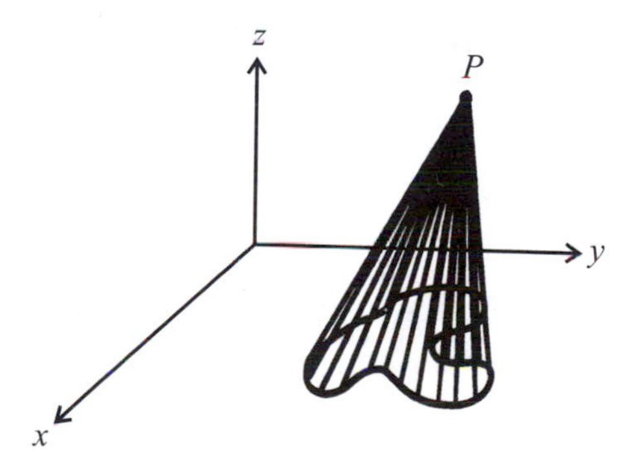

posed of a center point P and a dense set of spokes reaching out to the circular boundary, one can always create a circus tent construction by placing P in a third dimension to the plane. This obviates all possible self-intersections no matter how complex the shape of the curve. How then would you sew a disc to the boundary of a Möbius band?

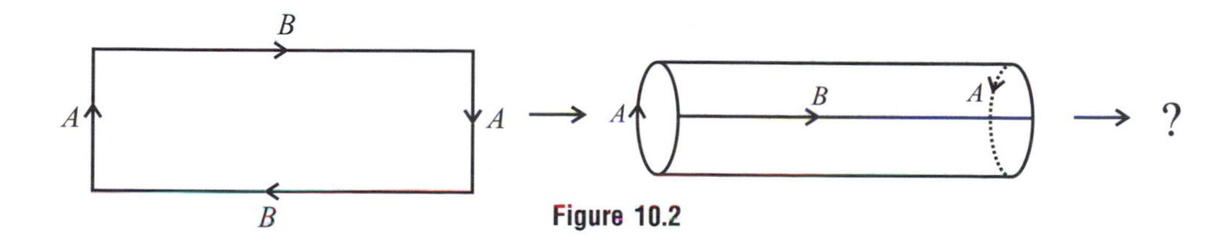

Figure 10.2

11 Paradoxes in Probability Theory

11.1 The Money or the Goat?

Try simulating this game with a friend using inverted plastic cups as doors and a piece of candy as the fabulous prize. Play the game several times, taking turns being the game show host and the contestant. What do you notice about the host's choices as to which "door" to reveal in the intermediate part of the game? How often does the contestant seem to win if a "never switch" strategy is adopted? What about an "always switch" approach?

11.2 Double or ... Double!!

There is a 50% chance that the other bag contains half as many Tootsie Rolls® as you just counted and a 50% chance it is double the number. By switching you could gain far more than you could lose. It is to your advantage then to switch. But wait! Wouldn't you follow the same line of reasoning no matter which bag you chose first? You would always opt to switch. Why then didn't you just take the other bag in the first place? What's going on?

11.3 Discord among the Chords

If you perform the experiment suggested, you will find that about half of the wires cross the circle at lengths greater than the side length of the triangle. Joi's reasoning is absolutely correct.

But there are other ways to select chords at random. What if you spin a bottle in the center of the circle to select points on the perimeter to connect with a chord, or throw a dart at a circle to determine the midpoint of a chord? Both Jennifer and Bill's lines of reasoning are absolutely correct, too! What's going on?

11.4 Alternative Dice

Yes, it is possible. Label one die with 1, 3, 4, 5, 6, and 8.

12 Don't Turn Around Just Once

12.1 Teacup Twists

This puzzle is interesting only if you have three or more strings attached to the cup. It is easy to untangle two strings no matter how many times the cup is rotated. Given just one rotation, it is impossible to untangle three or more strings. (Try the case with just three strings. Also see section 5.3.) But with two rotations this is no longer the case! It suddenly becomes possible to untangle all strings. Try lining the strings in a row above the teacup as shown in the photo on the facing page before rotating it 720°.

12.2 Rubber Bands and Pencils

Lay a large band of paper on a table top. How many loops can you make along this band with-

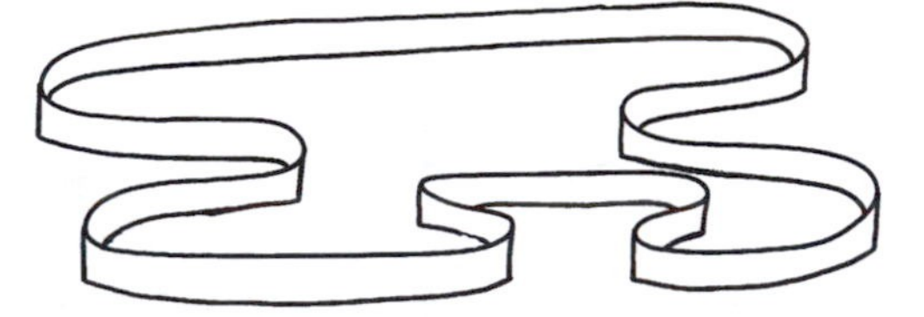

out producing any twists in the paper — that is, in such a way that along every part of the band the "wall" of paper remains essentially vertical?

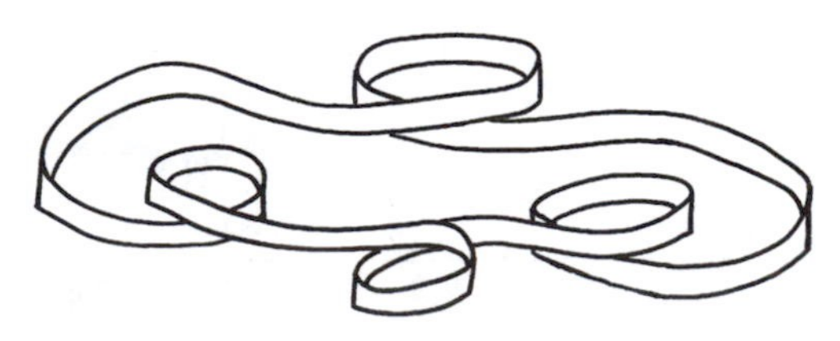

Diane Dixon and Lusine Ayrapetian hold up the strings before rotating the system 720°.

13 It's All in a Square

13.1 Square Maneuvers

When trying this out, after the fun of first letting everyone move simultaneously, try having one person move at a time, as follows. The first person moves to an occupied square. Its occupant then moves either to another occupied square or to an empty square—her choice. Continue in this fashion. Once someone moves to an empty square, a "circuit" of people is complete, and someone else must be nominated to take the next step. As you do this, count the number of people that constitute each circuit. What do you observe?

13.2 Path Walking

Color the grid according to the standard checkerboard coloring scheme. Any path you walk alternates between black and white cells. Is it possible then to commence a path on a white cell?

13.3 Square Folding

Does "63" sound familiar? What's going on?

14 Bagel Math

14.1 Slicing a Bagel

There is one other way to do this. Think diagonally!

14.2. Disproving the Obvious

Think of Christopher Columbus!

14.3. Housing on a Bagel

Alas, no! If there were a solution on the surface of a sphere, simply puncturing the sphere and stretching the surface flat would yield a planar solution. As no such planar solution exists (see [Char], Chapter 9 and [Gard15], Chapter 11) there can be no spherical solution.

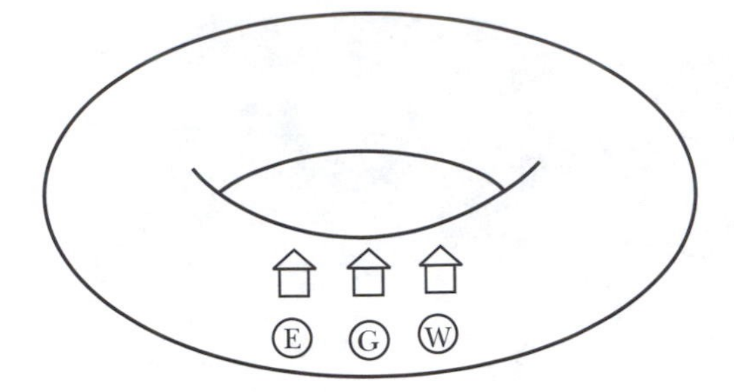

Taking It Further. What if the earth were the shape of a bagel? Could the problem be solved on a toroidal planet instead?

Alex Alapatt examines the housing problem on a bagel.

14.4. Tricky Triangulations

To answer the first question, count the number of edges in a theoretical triangulation by counting the number of triangles. (Each triangle has three edges. If you know that there are t triangles in all, do you now know the total number of edges?)

With regard to the second question, think of a torus as being formed from a square piece of paper by identifying (gluing) opposite edges. (See chapter 10.) Try experimenting with tri-

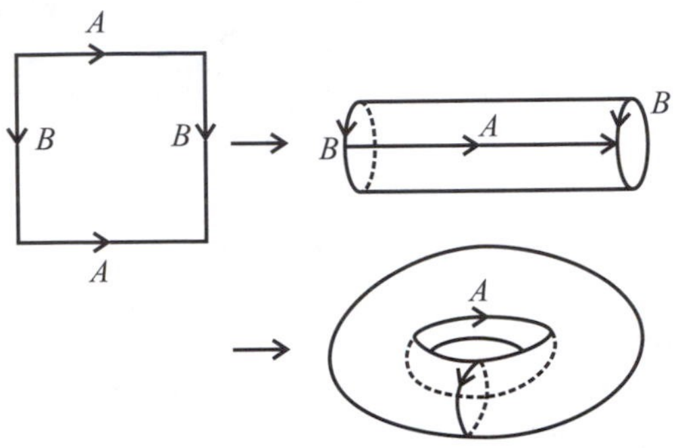

angulations on a square rather than the fully formed three-dimensional object. But keep in mind that in a valid triangulation of the torus the triangles can touch only at a single vertex or along a single edge. For instance, here are two pictures that *do not* represent valid triangulations of the torus.

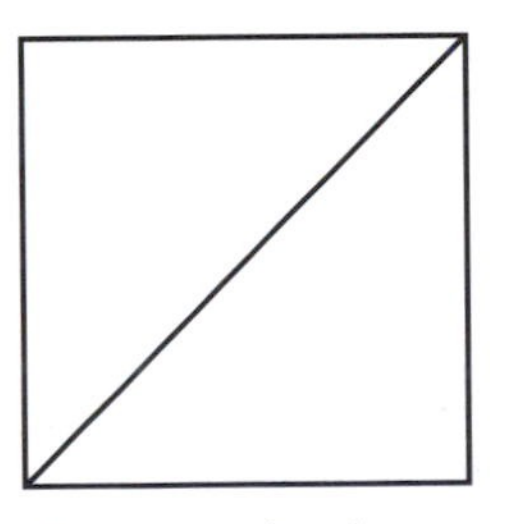

These two triangles touch along all three edges.

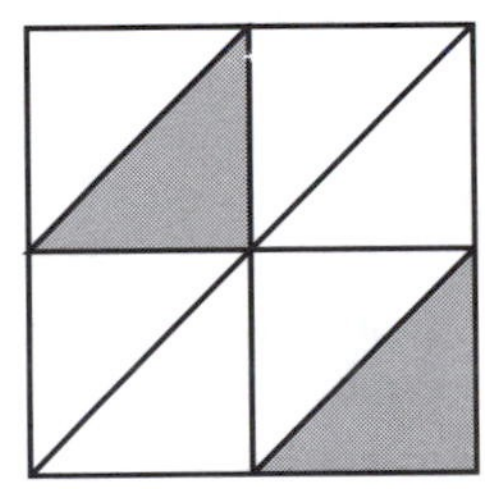

The two shaded triangles touch at two vertices.

The following drawing is, however, a triangulation of the torus. It uses 18 triangles — but you can do much better!

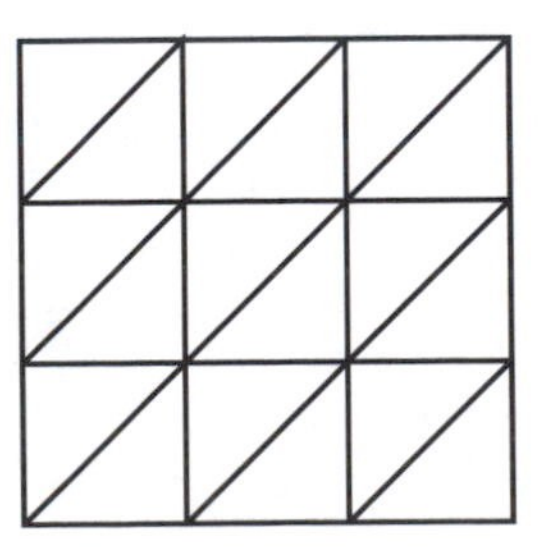

14.5. Platonic Bagels

Suppose m edges meet at each vertex. Count the number of edges in a theoretical picture in two different ways: If we are told there are a total of v vertices, do we now know the total number of edges? If we are told there are r regions, do we know the total number of edges?

15 Capturing Chaos

15.1 Feedback Frenzy

Why do you sometimes hear those horrible screeching sounds when people are setting up microphones and amplifying systems? The situation is analogous.

15.2 Creeping up on Chaos

With $r = 3.3$ the sequence no longer converges to a single value (as for $r = 2.5$). Rather a steady oscillation between two values results.

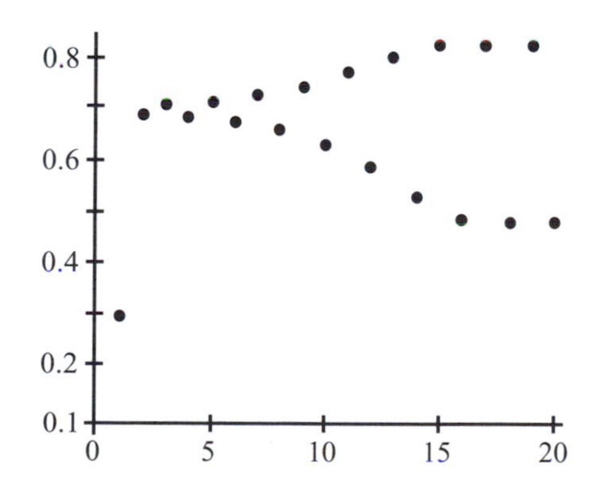

Challenge. By experimenting with intermediate values of r, can you determine precisely where this transition, or *bifurcation,* takes place? Keep going to see what happens with higher values of r.

16 Who has the Advantage?

16.1 A Fair Game?

Can you tell from experimenting who wins most often? Try to analyze mathematically a game where one player has three coins and the other two, and then one involving four and three coins.

16.2 Voting for Pizza

It really is in Alice's best interest to vote pepperoni. She reasons as follows: "Either Brad and Cassandra will agree and submit the same vote, or they will disagree in their choices. In the first case, it does not matter what I vote, for their choice will win. In the second case, however, pepperoni will win — either by a majority vote or by my veto power within a three-way tie. It is to my advantage then to vote pepperoni."

But Brad and Cassandra are aware that Alice must reason this way and she will vote pepperoni. Knowing Alice's vote, how could Brad and Cassandra now turn the situation to their advantage?

16.3 A Three Way Duel

Notice that Alberto survives about 31% of the time, Bridget 54% of the time, and Case about 15%. Repeat the experiment again, but this time have Alberto *deliberately miss* his first shot, killing no one even if his shot would have been successful. Have Bridget and Case play as before. What do you notice this time? What's going on?

16.4 Weird Dice

Notice die A will beat B two thirds of the time. How often will B beat C? C beat D? D beat A? If you choose a die first, which die do you think I will choose in response?

17 Laundry Math

17.1 Turning Clothes Inside Out

All clothes are basically the same shape: they are spheres; punctured spheres, to be precise. Socks are basically deformed spheres with the ankle holes providing punctures in their sides. Trousers, sweaters, and even button-down shirts are spheres with three punctures (ignoring the buttonholes). The eversion of such objects is straightforward: Everting a punctured sphere clearly yields another sphere, and the shape and structure of clothing thus does not change.

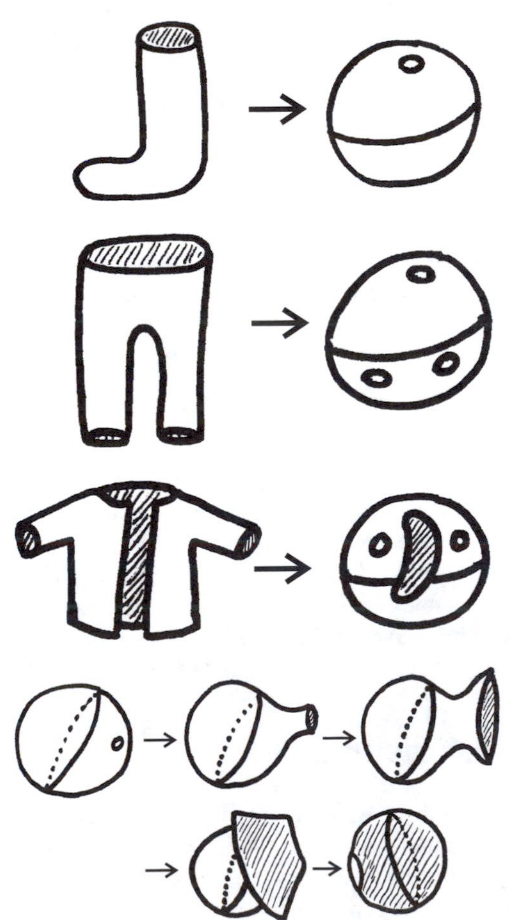

Taking it Further. Is it possible to evert a *non*-punctured sphere?

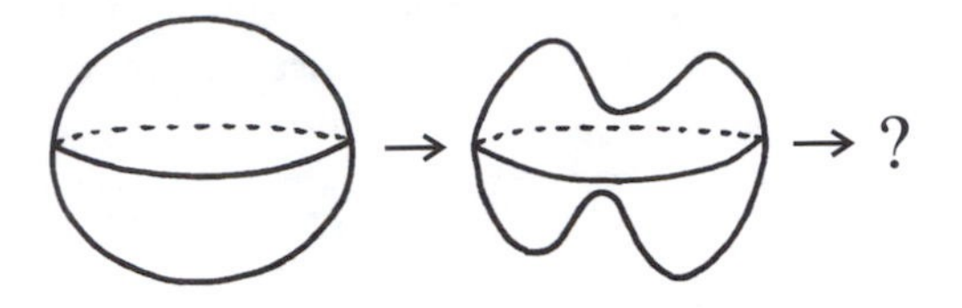

17.2 Mutilated Laundry

As we saw in 17.1, turning a sphere inside out yields another sphere. Surprisingly, turning a punctured donut inside out yields another donut! What happens for multi-donuts?

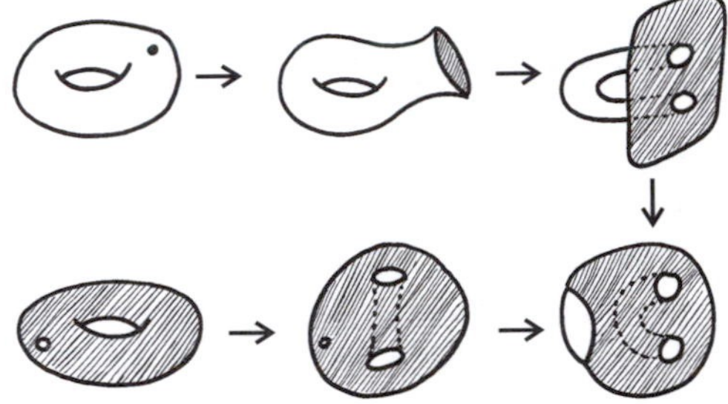

Taking it Further. Notice that the material forming the tube or *handle* of the original donut before eversion becomes the material forming the hole of the donut after eversion. For a pair of trousers this is significant. Two trouser legs sewn together form a very long and skinny donut handle. Upon eversion this must become the donut hole and the result is an elongated donut difficult to recognize as such at first. (Did you see it as another donut when you tried it? Perhaps try instead everting a punctured donut sewn together from a square piece of material.)

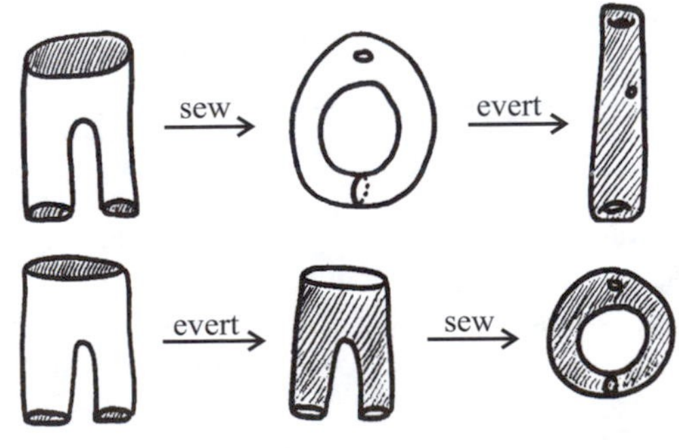

The surprising thing is that the outcome is *not* the same as first turning a pair of trousers inside out and then sewing the two legs together. The actual eversion produces an alternative result. We have discovered then two, possibly different, eversions of the same donut! To what extent are these two eversions distinct? Is there some way to manipulate one eversion into the other with the same pair of trousers? Try experimenting with a pair of trousers again.

17.3 Cannibalistic Clothing

Try it! See what happens.

18 Get Knotted

18.1 Party Trick I: Two Linked Rings?

I have deliberately misled you in my note! It is imperative that the strings be tied around the participants' wrists rather than held by hand if they are to escape.

18.2 Party Trick II: A T-Shirt Trick

In both cases it is possible!

18.3 Party Trick III: A Waistcoat Trick

Begin by slipping both arms through the arm holes of the waistcoat.

18.4 Two More Linked Rings?

Begin by molding the stem of one loop across the body of Play-Doh® and then to the base of the other ring. (In terms of a human figure this would mean molding the base of a thumb down the length of one arm, across the chest and then up the other arm.)

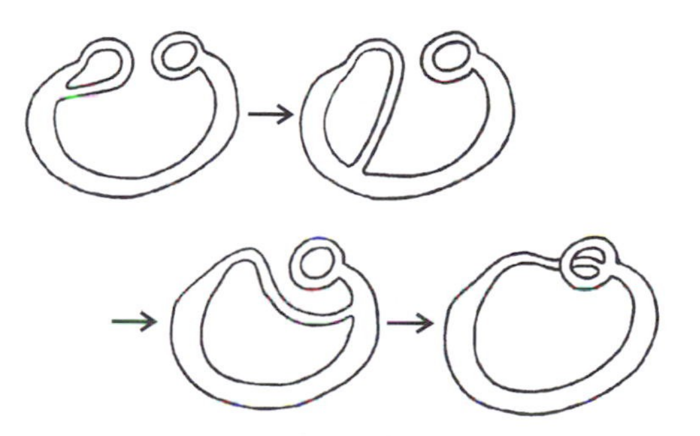

19 Tiling and Walking

19.1 Skew Tetrominoes

There is essentially only one way to place a skew tetromino in the topleftmost corner of a square or rectangular grid while staying within the boundary of the grid. The placement of this tile then forces the placement of a tile immediately below it, which in turn forces the placement of another tile below that, and so on, all the way down to the bottom leftmost corner. Assuming we want to stay within the boundary of the rectangle (which we do!), there is no room to place the final tile at the end of this chain. Thus it is impossible to tile any size rectangular grid with skew tetrominoes.

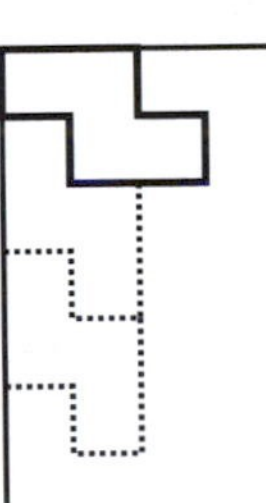

Taking it Further. Is it possible to tile this region with skew tetrominoes?

19.2 Map Walking

Such paths do exist. Given the inherent symmetry in the diagrams, none of them depend on the choice of starting intersection for either participant.

19.3 Bringing it Together

Place a copy of the boundary of the region on the map for city A. In walking the path of this boundary, will the inhabitant of city B, following a shadow journey, also walk a closed loop? What does this tell us about the tilability of the region?

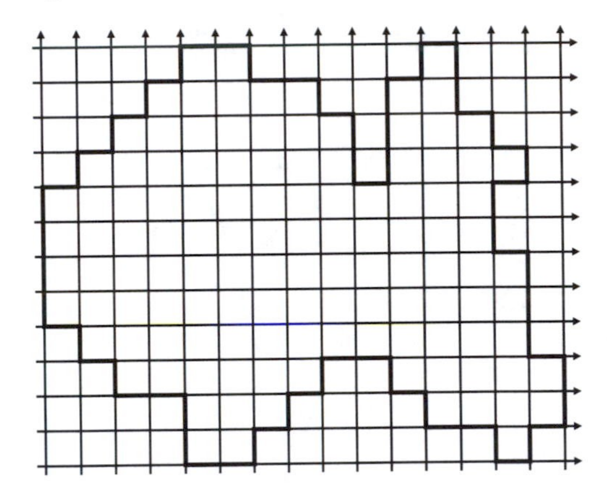

20 Automata Antics

20.1 Basic Ant Walking

How many steps must the ant take to complete a loop?

20.2 Ant Antics

A 7×7 grid is a bit unwieldy. Try the problem on a 3×3 grid instead.

20.3 Ball Throwing

We can view this problem as one of ant motion on a *complete graph*. Such a graph consists of finitely many vertices with edges connecting all possible pairs of vertices. Each vertex is labelled "L" or "R" and the ant wanders from vertex to vertex turning either the sharpest possible left or sharpest possible right according to the label of the vertex it visits (and then changing the label of that vertex). Notice that the sharpest possible left turn could actually be a right turn if the ant is travelling along an outer edge approaching a vertex labelled L (hence our peculiar convention for the special case of ball throwing).

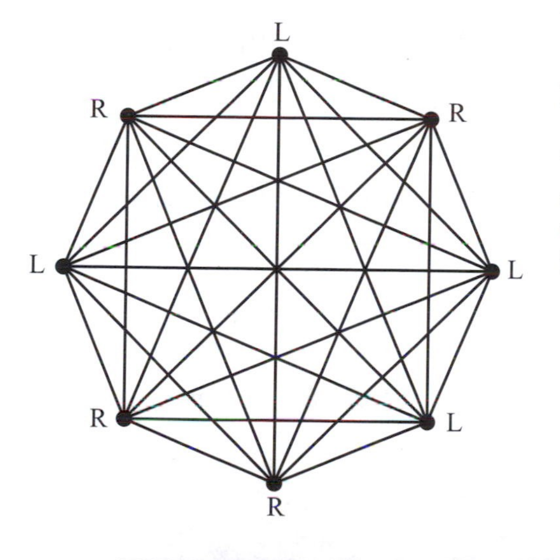

Another way to experiment with this ball throwing problem is to draw a large complete graph on the floor and have one person move along its edges like the ant. People standing at the vertices help direct the ant's motion by calling out the appropriate left or right turn. Note that the role of L's and R's is reversed in this interpretation.

21 Bubble Trouble

21.1 Road Building

Try a symmetrical design as shown. If the towns are situated on a square one (large) unit wide, what value for x is best? (*Warning:* Calculus is needed!)

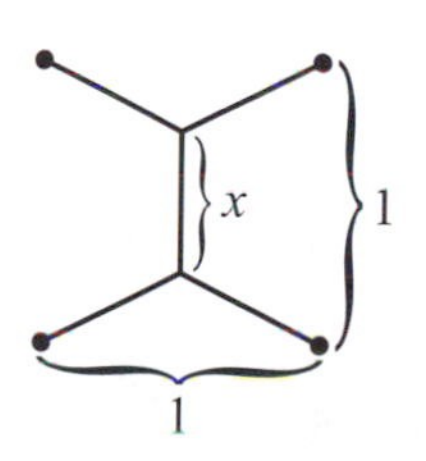

Merrimack College students and faculty perform the ant walking experiment.

21.2 Higher Dimensional "Road Building"

Don't be afraid to dip your elbows into soap! Try it!

21.3 Donut Bubbles

Actually, no! It has long been known [Hild] that the sphere is the surface of least area that traps a given amount of air. Any donut bubble you create (did you succeed, even for an instant?) will wobble, deform, and likely burst before pulling itself into a sphere. However, is it possible to create a donut bubble as part of a double bubble? Think about a donut bubble that enwraps a single dumbbell shaped bubble.

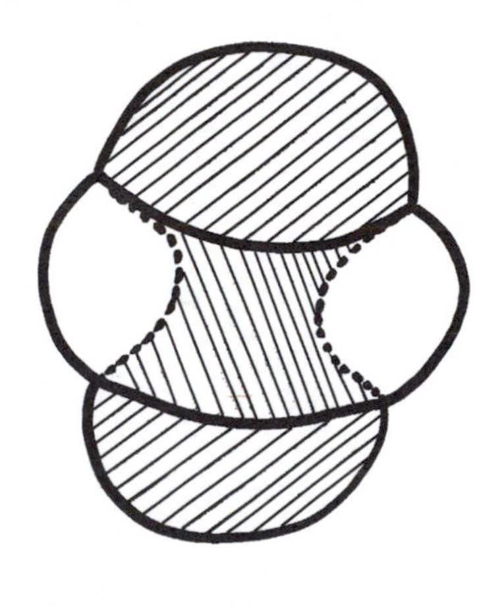

22 Halves and Doubles

22.1 Freaky Wheels I

The arrows are again aligned and pointing upwards! Can we conclude that rolling one wheel halfway along the circumference of another is the same as rolling it along the whole? What's going on?

22.2 Freaky Wheels II

Have you ever parked a car too close to the curb? The terrible screeching sound you hear is the hubcap of your back wheel scraping against the curb. The tire rolls, but the hubcap scrapes!

22.3 Breaking a Necklace

No matter how the pearls are arranged, only two cuts are ever needed, and these cuts can always be placed at opposite points on the circle! Try the experiment several times. Can you begin to explain why this is always the case?

22.4 Congruent Halves

They all can!

23 Playing with Playing Cards

23.1 A Pastiche of Card Surprises

Surprise 1. They are always equal!

Surprise 2: Six! (Is there anything special about the numbers 32 and 20?)

Surprise 3: The number of foreign cards in each pile is always the same! (Does anything special about the number ten make this work?)

Surprise 4: The card you first noted!

Surprise 5: Each pile consists of the same numbered cards!

Surprise 6: The magic card!

Surprise 7: Your friend will be handed the card he mentally selected!

Why do these tricks work?

23.2 Curious Piles

No hints, here. Instead …

Taking It Further. Suppose a shuffled deck of cards is divided into 13 piles of four cards each. Is it always possible to select an ace from one pile, a two from another, a three from a third and so on, all the way down to a king from a thirteenth pile? That is, can you select 13 distinct numbered cards from 13 distinct piles? Try it and see if you can accomplish this feat. Or, try to produce an arrangement of cards for which this cannot be done.

23.3. On Perfect Shuffling

Denote an in-shuffle by I and an out-shuffle by O and read a string of these letters from left to right as a sequence of instructions. Thus I I O, for example, means to perform two in-shuffles followed by an out-shuffle. It is convenient to regard the top position of the deck as position 0, the next card down as position 1, and so on.

Make a table of the sequence of steps required to move a card in position 0 to another position. (I have completed the first three entries for you.) Is there any connection here to binary numbers?

Destination	Moves
1	I
2	I O
3	I I
4	
5	
6	
7	

24 Map Mechanics

24.1 Cartographer's Wisdom

I forgot to mention that I wanted the outside region of the map, that is, the border between the map and the edge of the page, colored as well. Does this disrupt your coloring scheme?

24.2 Simple Map

Ignoring the outside region, two colors suffice to color this map.

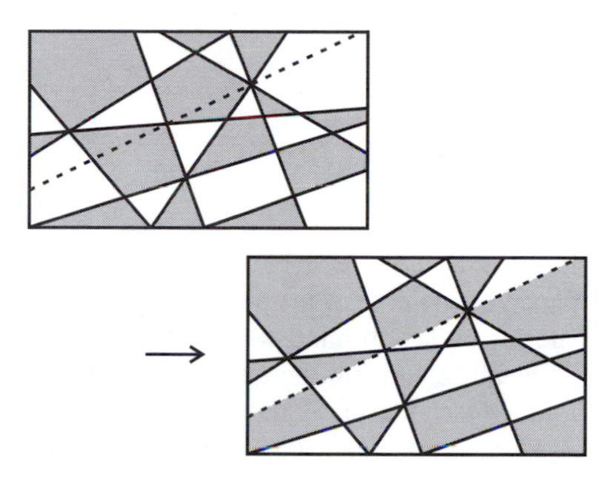

Taking it Further Answer. When adding an extra line to the diagram, just reverse the colors of the regions to one side of it. This produces a satisfactory two-coloring of the new map.

Comment. This trick works as the basis of an induction argument to prove that all maps constructed from straight lines drawn across a page are two-colorable. The same trick also works when slicing a large three-dimensional cube by planes, subdividing its interior into separate regions: it is always possible to fill the volumes of these regions with red and blue liquids so that no two regions sharing a common face are assigned the same colored liquid.

Taking it Even Further. A single curly line is drawn from one end of a page to the other. The line intersects itself at isolated points but not along entire segments of the curve. Is the resulting map necessarily two-colorable?

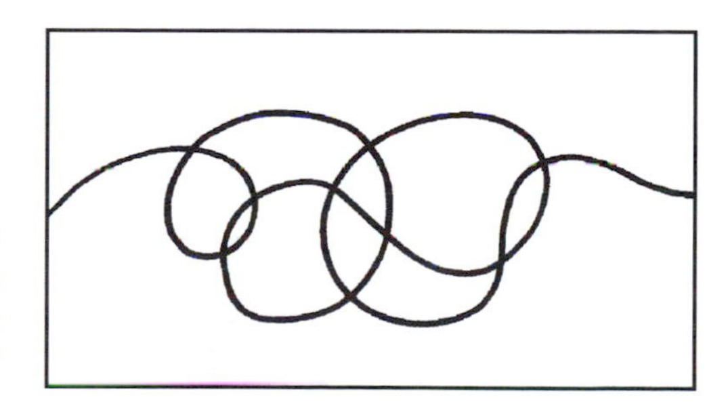

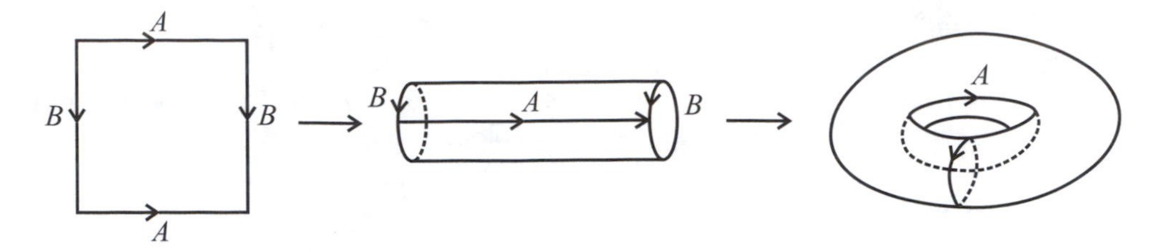

24.3 Toroidal Maps

Rather than work with a fully formed torus, it is easier to draw maps on a square piece of paper, taking note of the proper *edge identifications.* (When forming the torus, the top edge is actually glued to the bottom, and the left edge to the right.) See the figure above.

Here's a toroidal map requiring a minimum of five colors to paint.

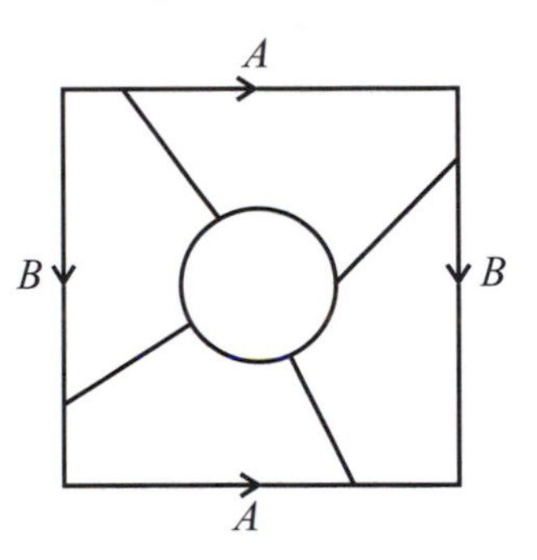

Taking it Further. Design a map on a torus that requires a minimum of *six* colors to paint. How about one requiring a minimum of *seven*?

25 Weird Lotteries

25.1 Winning Cake

A number of possible strategies are worth considering. Would you choose a large number in order to win, or choose a small one and hope that the entries above you cancel out? Would you want to announce to the group what number you are going to write? Would you want to announce a number but secretly write a different one instead? Or is it best to keep silent? Does the size of the group affect your strategy? What if only 2, 3, or 4 people are with you?

The best thing is to try running this game and see how people operate. There is a clear trade-off, and it is interesting to see how people respond to it.

Taking it Further. Suppose you are allowed to hand in as many entries as you like. What will you do in this case?

Taking it Even Further. Suppose you are told that after the winner takes his or her share, the rest of the cake will be divided equally among the remaining participants. Will you now adopt a different strategy?

25.2 Unexpected Winner

If the professor is truthful, John has indeed won the cake. But then the announcement of the winner wouldn't be a surprise, contradicting the assumption that the professor told the truth! Has the professor lied?

25.3 Winning Tootsie Rolls®

It is hard to resist the temptation to defect. Do you think in a small group everyone would be likely to cooperate? Or would everyone defect?

Taking it Further. Try this variant scheme. If everyone cooperates, everyone receives five Tootsie Rolls® apiece. If everyone defects, no-one receives any Tootsie Rolls®. If there is a mixture of cooperators and defectors, the co-operators each receive two Tootsie Rolls® and the defectors twenty. What is your response?

Outcomes	C	D
All Cooperate	5	–
All Defect	–	0
Mixture	2	20

25.4 Buying Tootsie Rolls®

Many people feel that investing in the stock market is like entering into a lottery! This financial problem is modelled on the mathematical idea of *dollar averaging*. Is it better, in the long run, to invest a fixed amount of money every month in a stock or to buy a fixed number of shares every month? If the price variations are truly random, one method is indeed better than the other. Can you detect which is better by experimenting?

26 Flipped Out

26.1 A Real Cliff-Hanger

After simulating the experiment a few times you might start to feel that things don't look good for Dorothy.

26.2 Too Big a Difference

After a large number of trials, the average difference between the number of heads and tails is about 2.46. Of course this average difference is *not* zero since I asked you to record all differences as positive quantities. (If, however, the appearance of more heads than tails is recorded as a positive difference, and the appearance of more tails as heads as a negative difference, then the average value of these differences would be zero.)

Repeat the experiment, again performing 10 tosses many times, but this time record the square of the difference. (In an experiment yielding three heads and seven tails, for example, the square of the difference is $4^2 = 16$.) Compute the average value of the square of the difference. What do you notice?

The result is even more startling if you repeat the exercise by tossing a coin 25 times.

26.3 A Surprise

Count the number of heads and tails that appear. Repeat the experiment several times if nothing striking occurs right away.

27 Parts That Do Not Add Up to Their Whole

27.1 A Fibonacci Mismatch

Try doing the same trick to transform a 5 × 5 square into a 3 × 8 rectangle (that is, choose $F_5 = 5$ to begin with). Make a 3 × 3 square into a 2 × 5 rectangle. What do you observe?

27.2 Cake Please

Have each brother place a knife, in parallel directions, across the cake, at the position each believes divides the cake in two. It is unlikely both will choose exactly the same line. (Use extraordinarily thin knives!) Cutting the cake anywhere between these two lines guarantees each brother a piece, in his estimation, greater than half.

This method generalizes to more than two participants. For instance, it is possible to divide a cake among seven people so that each person believes he is receiving more than one seventh of the cake! First have each person mark a line, parallel to one fixed direction, that he believes cuts off exactly one seventh of the cake from the left. Then make a cut between

cut
Albert
Hubert
Albert's piece
Hubert's piece

the two leftmost lines and hand that piece to the person who marked the line closest to the end. This person is receiving more than one seventh of the cake in his estimation, and the remaining six people all believe that more than six sevenths of the cake remains. These six folk now repeat this procedure, each estimating one sixth of this portion of cake, and so on, reducing the problem to smaller numbers of people until eventually every one ends up with a piece of the cake.

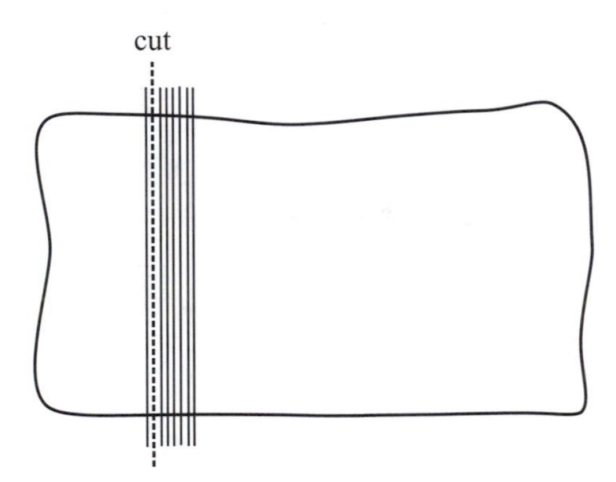

In case you are wondering, it is not to anyone's advantage to be greedy in this procedure and deliberately mark a line larger than the estimated one seventh. That risks receiving too small a piece in the end!

Comment. It is fun to try this experiment with a group of friends. Draw on a piece of paper a curved figure roughly resembling a rectangle. Make photocopies of this shape. Paper "cake" has the advantage that bold lines representing knife cuts can be drawn and seen when copies are stacked on top of one another. This way people can draw knife lines free of the visual clutter of other people's score marks.

Hand copies to groups of two. See if they can devise their own cake-sharing methods. Next hand out copies to groups of three and see what schemes they devise for fair distribution. There are several techniques for sharing cake and it is interesting to see the varied methods different groups devise.

Warning. If you illustrate these sharing techniques over real cake, be careful to choose a cake that is plain and free of decoration. Battles often ensue over iced flowers, for example, and little hope remains for rational mathematical analysis!

Taking it Further. The cake-sharing method described above is said to be *proportional*: everyone receives in her estimation at least $1/n$ of the cake (if there are n people). But it is not *envy-free*: Not everyone is likely to feel that she actually received the largest piece cut! (Except in the $n = 2$ case.)

Devise a method among three players that guarantees each participant, in her estimation, the largest (or at least tied for largest) piece ever cut.

27.3 Sharing Indivisible Goods

Bjorn should keep the bar and give Elaina 16 Tootsie Rolls®. Both come out the equivalent of two Tootsie Rolls® ahead!

Taking It Further. Neal, Janice, and Sheryl, each possessing a large supply of Tootsie Rolls®, are faced with the challenge of sharing four desserts: an apricot dacquoise, a hazelnut torte, an Australian pavlova, and a single American brownie. Each person agrees the desserts should not be cut in any way, so they can take a dessert home intact to share with the family, but each is willing to trade Tootsie Rolls® for desserts. They make the bids shown in the table:

Dessert	Neal	Janice	Sheryl
Dacquoise	105	120	132
Torte	90	80	64
Pavlova	196	75	112
Brownie	5	7	4

Who should take home which dessert? Who should trade Tootsie Rolls® and how many? Devise a scheme so that everyone, in his or her measure, comes out with more than one third of the share.

28 Making the Sacrifice

28.1 The Josephus Flavius Story

Number the participants 1 through n in a clockwise direction. Assume a game begins the count with player 1 and goes in a clockwise direction. Let $W(n)$ denote the position of the winner in an n-person game (counting every third person). The following table shows the position of the winner for the games involving $n = 1$ through $n = 15$ people:

n	1	2	3	4	5	6	7	8	9	10	11	12	13	14	15
$W(n)$	1	2	2	1	4	1	4	7	1	4	7	10	13	2	5

What do you notice about these numbers?

28.2 Soldiers in the Desert

Configuration A (below) will move a penny three lines into the desert. Configuration B (below) will push a scout four lines into the desert. Is it possible to push a penny five lines in?

28.3 Democratic Pirates

In a game with ten pirates, the cabin boy can survive and end up with six coins! Can you see how?

29 Problems in Parity

29.1 Magic Triangles

The instructions become more restrictive as you follow them, so it is hardly a surprise that I

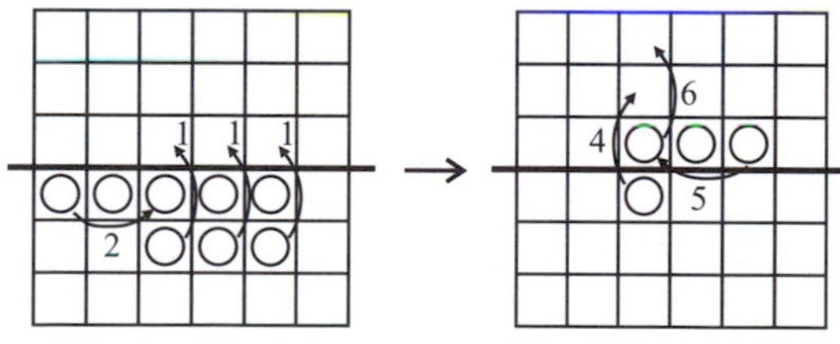

Configuration A. Moving a penny three lines into the desert.

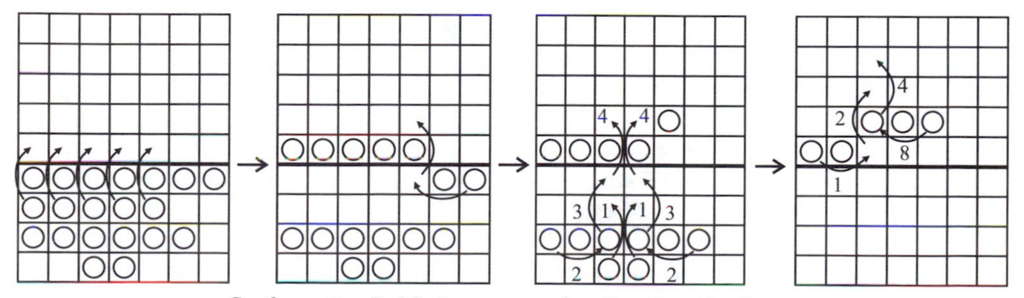

Configuration B. Moving a penny four lines into the desert.

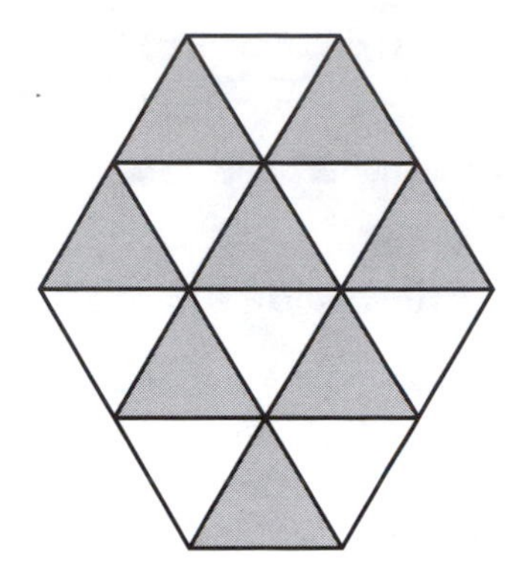

was able to squeeze your available options down to the one choice of moving into cell 23. To check this, try crossing out all the disqualified cells as you proceed through the instructions. However, my certainty that you avoided all inappropriate cells (all multiples of 3 after step 1, all multiples of 13 after step 5, for example) took a little more subtlety. I relied on the above picture. Do you see how?

29.2 Walking a Loop.

Consider the journey I described. Write out the letters alternately above and below a horizontal line, ignoring the final A (it was already mentioned as the starting point of the journey):

A B C J K D G F I L H E

D J H G I A L K F C E B

What do you notice?

29.3 Catch Me If You Can

Joy had better head to Port Moresby right away before doing anything else. Can you see why?

29.4 A Game of Solitaire

First try playing a game with a line of just one, two, or three pennies. Then try one with four pennies. Under what circumstances can you win the game? What do you notice about the number of heads showing initially for these winning games?

30 Chessboard Maneuvers

30.1 Grid Walking

Color the cells of the grid black and white according to a standard checkerboard scheme.

Each step Greta or Peter takes leads to a cell of opposite color. Thus an even number of color changes (that is, an even number of steps) is required to visit a cell of the initial color. In particular, an even number of steps is required to return to the initial cell. What can be said for Lashana's and Lopsided Charlie's motion?

30.2 Kingly Maneuvers

After trying a number of times (or playing the two-person game for a while), you probably suspect that it is impossible to avoid simultaneously the creation of such black and white paths. Your feeling is correct. To see why, imagine the puzzle as one of trying to dam a river. View the black cells as logs attempting to link across a whitewater river to block the flow.

30.3 Mutual Non-Attack: Rooks

The placement of a rook in a particular cell of an $n \times n$ grid "knocks out" all cells in that row and column as possible placements of other rooks. This leaves an $(n-1) \times (n-1)$ grid to analyze. Placing one more rook leaves an $(n-2) \times (n-2)$ grid to analyze, and so on. This shows that a maximum of n rooks can be placed on the board.

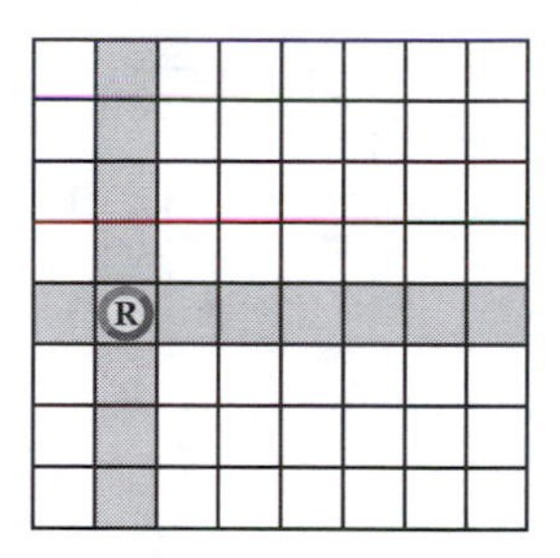

Note that each row and column of the grid must contain precisely one rook. When construct-

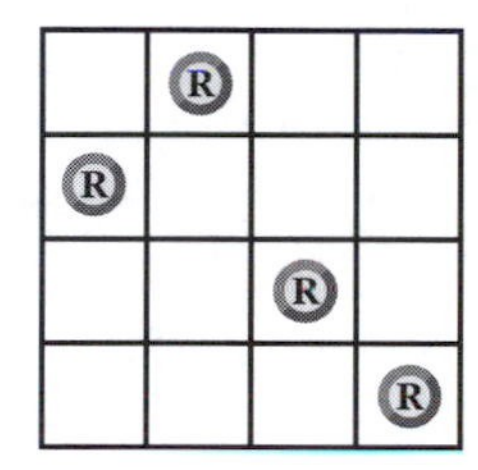

ing a configuration of rooks, there are n choices for where to place a rook in the first row. Once this first rook is in place, there are $n-1$ choices for where to place a rook in the second row, and so on. This gives a total of

$$n \times (n-1) \times \cdots \times 2 \times 1 = n!$$

possible arrangements of n rooks on an $n \times n$ chessboard.

For a $p \times q$ rectangular array, with $p < q$, only p rooks can be placed on the board. The "one rook per row" rule dictates this. There are $q \times (q-1) \times \cdots \times (q-p+1) = q!/(q-p)!$ ways to arrange these rooks.

Taking It Further 1: Toroidal Chessboards.

One can form a torus from a square piece of paper by gluing the top edge of the square to the bottom edge to first form a cylinder, and then the left edge to the right to form a torus (see the figure below).

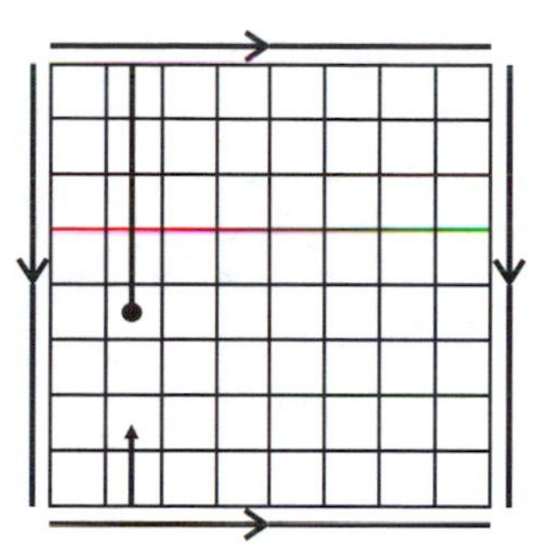

Imagine that an 8×8 chessboard has been folded to form a torus. Still regard the board as a planar square grid, but keep in mind that it is possible to move beyond the top edge to reappear at the corresponding neighboring cell on the bottom edge. The same is true for the left and right edges. What is the maximal number of rooks that can be placed on an 8×8 toroidal chessboard in mutual non-attack? Given this maximal number, what is the total number of different configurations possible?

Taking It Further 2: Möbius Chessboards.

This time glue just the left edge of an 8×8 checkerboard to the right edge, but insert a half twist. This forms a Möbius chessboard. What is the maximal number of rooks that can be placed on this twisted board in mutual non-attack? Given this maximal number, what is the total number of different configurations possible?

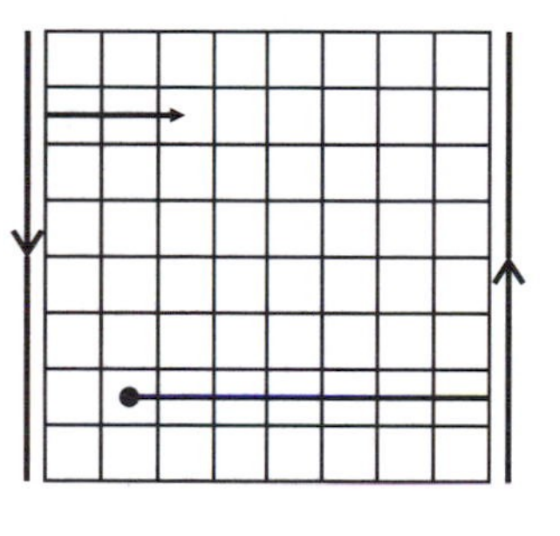

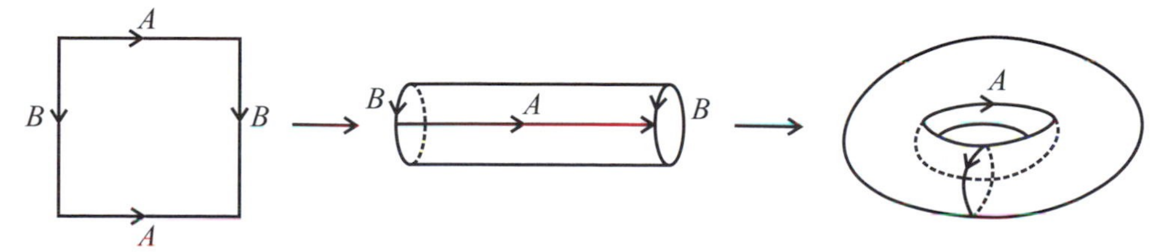

Taking It Further 3. Projective Plane Chessboards. Suppose the left and right edges, and the top and bottom edges, of an 8×8 chessboard have been glued together with a half twist. This forms a surface, difficult to draw, called a *projective plane* (see section 10.2). What is the maximal number of rooks that can be placed on this board in mutual non-attack? In how many different ways can they be placed?

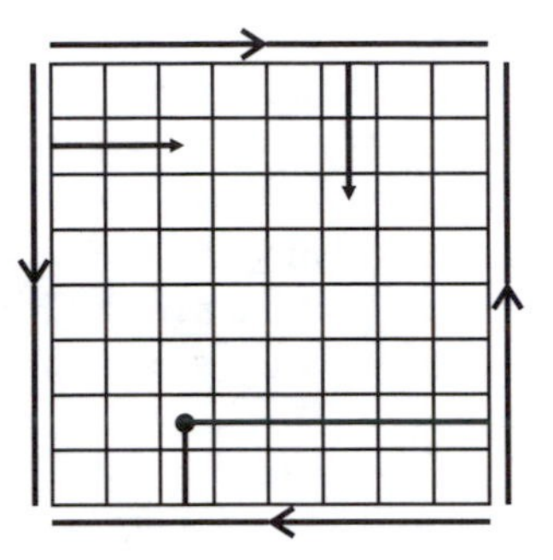

Challenge. Develop solutions to the previous three problems for arbitrary $p \times q$ chessboards!

Taking It Further 4: Three-Dimensional Chessboards. Imagine a rook standing within one cell of a cubical lattice. Assume it can move

in straight lines up, down, left, right, and back or forth to attack other pieces. In a $3 \times 3 \times 3$ cubical lattice, what is the maximal number of rooks that can be placed in a configuration of mutual non-attack? In how many different ways can they be placed? Can you generalize this analysis to $n \times n \times n$ cubical lattices?

30.4 Mutual Non-Attack: Queens

Given what we know about rooks, it is not possible to place more than n queens on an $n \times n$ chessboard in mutual non-attack. It is possible to place four queens on a 4×4 board, five on a 5×5 board and eight on an 8×8 board. (Did you succeed with this?) Do you think this pattern persists?

PART III

SOLUTIONS AND DISCUSSIONS

1 Distribution Dilemmas

1.1 A Shepherd and His Sheep

The sons should borrow one sheep from the neighboring daughters to boost their flock to a total of 18 sheep. From this, the first son receives nine sheep, the second six, and the third two. This leaves one sheep left over to be returned to the daughters!

Taking It Further Answer. The three daughters can accomplish a similar feat—if they have the means to borrow 1650 sheep from another source! This would boost their flock to 2145 from which the first daughter receives 429 sheep (one fifth), the second 65 sheep (one 33rd), and the third one sheep (one 2145th!). This leaves 1650 sheep, which can be returned to their original owners.

Question: Are these solutions legally satisfactory? May an estate accrue debt *before* the distribution of goods?

It is easy to create similar problems of this type. Simply collect a list of favorite fractions (that sum to less than 1) and represent their sum as a fraction of the denominators' lowest common multiple. For example, the sum

$$\frac{1}{5}+\frac{1}{6}+\frac{5}{12}+\frac{3}{20}=\frac{56}{60}$$

yields the puzzle: "Four sons must divide an estate of 56 sheep according to the proportions $^1/_5$, $^1/_6$, $^5/_{12}$, and $^3/_{20}$."

Those puzzles involving *unit fraction* proportions (fractions with unit numerator) and the transfer of one sheep for their solution are particularly appealing. The equations,

$$\frac{1}{2}+\frac{1}{3}+\frac{1}{9}=\frac{17}{18}$$

$$\frac{1}{2}+\frac{1}{3}+\frac{1}{10}+\frac{1}{20}=\frac{59}{60}$$

$$\frac{1}{2}+\frac{1}{3}+\frac{1}{4}+\frac{1}{6}+\frac{1}{8}+\frac{1}{12}=\frac{23}{24}$$

for example, yield such puzzles.

By the way: Four lines can indeed connect nine dots.

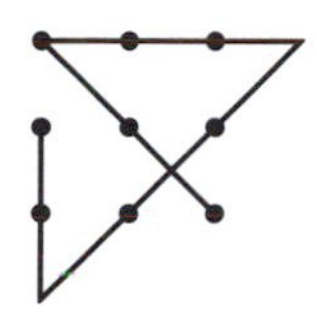

Challenge. Connect a 4×4 array of dots with just six contiguous straight line segments.

Hard Challenge. Is there a formula for the minimal number of contiguous straight line segments required to connect an $n \times n$ array of dots? How about for rectangular arrays of dots?

A Note on Unit Fractions

Unit fractions have a prominent role in the history of mathematics. The Egyptians of 4000 years ago felt the need to express all fractional quantities as sums of *distinct* unit fractions. For example, $^3/_{10}$ was written $^1/_4 + ^1/_{20}$, and $^5/_7$ as $^1/_2 + ^1/_5 + ^1/_{70}$. Their methods for computing these unit fraction expansions were ingenious, but this notational system is, to say the least, cumbersome. Many historians believe it held back the Egyptians from making significant progress in their mathematics. Can *every* fraction be expressed as a finite sum of distinct unit fractions? It is not clear whether the Egyptians ever questioned this.

In 1202 Fibonacci began his own investigation of unit fractions and proved that every rational does indeed have a unit fraction representation. His proof is constructive in that it also tells you how to find particular unit fraction representations. Given a fraction $^a/_b$, select the unit fraction $^1/_n$ less than $^a/_b$ with n as small as possible. (Thus $^1/_n \leq {}^a/_b$ but $^1/_{n-1} > {}^a/_b$). Then

$$\frac{a}{b}-\frac{1}{n}=\frac{na-b}{bn}$$

is a new fraction with numerator smaller than a. (Check this!) Repeating this procedure produces another unit fraction $1/m$ such that

$$\frac{a}{b}-\frac{1}{n}-\frac{1}{m},$$

when expressed as a single fraction, has an even smaller numerator. A finite number of applications of this procedure must eventually produce a fraction with a unit numerator (and the unit fraction expansion of a/b is now apparent).

For example, applying this procedure to the fraction $4/13$ yields

$$\frac{4}{13}-\frac{1}{4}=\frac{3}{52} \qquad \left(\tfrac{1}{3} \text{ is too large a fraction to subtract}\right)$$

$$\frac{3}{52}-\frac{1}{18}=\frac{1}{468} \qquad \left(\tfrac{1}{17} \text{ is too large a fraction to subtract}\right)$$

and so

$$\frac{4}{13}=\frac{1}{4}+\frac{1}{18}+\frac{1}{468}.$$

Question: Is this expression unique? Can $4/13$ be expressed as a different sum of unit fractions?

A number is said to be *perfect* if it equals the sum of its proper divisors plus one. As

$$6=1+2+3,$$
$$28=1+2+4+7+14, \text{ and}$$
$$496=1+2+4+8+16+31+62+124+248,$$

6, 28, and 496 are perfect. Perfect numbers lead to particularly nice unit fraction expansions, with denominators being the divisors themselves. For instance,

$$\frac{1}{2}+\frac{1}{3}=\frac{5}{6},$$
$$\frac{1}{2}+\frac{1}{4}+\frac{1}{7}+\frac{1}{14}=\frac{27}{28},$$
$$\frac{1}{2}+\frac{1}{4}+\frac{1}{8}+\frac{1}{16}+\frac{1}{31}+\frac{1}{62}+\frac{1}{124}+\frac{1}{248}=\frac{495}{496}.$$

This pattern holds for all perfect numbers. Can you see why?

1.2 Iterated Sharing

No matter the initial distribution of candy or how many people are in the circle, the distribution of candy will eventually equalize: Everyone will end up with the same amount! This can be seen as follows.

Suppose the most candy any individual possesses is M pieces. Upon sharing, such an individual gives $M/2$ pieces to her neighbor on the left and receives no more than $M/2$ pieces from the right (since M is the maximal number of pieces anyone possesses). She ends up with no more than M pieces as a result. She will then be given an extra piece of candy only if her resultant pile is an odd number. But since M is even, possession of $M+1$ pieces will never result. Thus the number of pieces any individual can ever possess is bounded above by the maximal amount M given in the initial distribution.

Now consider those folks possessing the smallest amount of candy. Let n represent this minimal amount. Upon sharing, such an individual will pass $n/2$ pieces to his left neighbor and receive at the very least $n/2$ pieces from his right. Only if the person to his right possesses the same minimal amount n of candy pieces will his pile fail to grow, otherwise he will end up with $n + 1$ pieces or more (and maybe even the addition of an extra piece!). Thus if there are k people sitting in a row, each with n pieces, the first $k - 1$ people will stay at n pieces and the kth person will end up with more (unless *everyone* has n pieces, in which case the iteration has stabilized.) Thus the number of people with a minimal amount of candy decreases on each iteration. Given the upper bound on the amount of candy one can possess, this fact forces the situation to stabilize!

Taking It Further Answer. Eating the odd pieces of candy between iterations also induces an equal distribution in the long run. The reasoning is essentially the same as above, except that there is a lower bound n on the amount of candy an individual can possess as the number of people with a maximal amount of candy decreases.

Sharing all your candy with both your left and right neighbors can produce oscillatory patterns if the number of people in the circle is even. Imagine folks changing seats between iterations, moving one seat to the right to follow their right half-pile. The situation is thus equivalent to each person simply giving half of his candy to the person two places to his left. With an even number of players, this breaks the analysis into two disjoint groups, each of which must stabilize (but not necessarily to the same number). For an odd number of people, the problem is equivalent to just one circle, which must stabilize.

There are a myriad of ways to explore this problem further. What happens if thirds or quarters of piles of candy are shared with two or three people? What if candy is passed to different positions along the circle? What if different people pass candy to different places—say, the first person one place to the left, the second two to the left, and so on. Must the situation always stabilize?

Returning to the original problem, given the number of people in the circle and the initial distribution of candy, is it possible to predict how much candy everyone will end up with in the end? Can you predict how many iterations it will take to reach the stable pattern? The students at St. Mary's College of Maryland have explored this question and have found some interesting patterns.

Acknowledgments and Further Reading

A wonderful article by W. G. Frederick and J. Hersberger [Fred] discusses the classic inheritance problem and a host of related problems with amusing twists. Unit fraction expansions are still an area of active research; see the article by C. Anne [Anne], which also mentions a website created by D. Eppstein, containing a great deal more information about unit fraction expansions and efficient algorithms for computing them rapidly. For a brief overview of the work of Egyptian mathematicians see chapter 0 of William Dunham's wonderful book [Dunh]. The first variant of the candy distribution problem appears in the marvellous work [Chan] by Gengzhe Chang and Thomas Sederberg. Have a look at their chapter 7 for a non-discrete extension of the puzzle.

2 Weird Shapes

2.1 Plucky Perimeters

With the appropriate enlargement or reduction it is theoretically possible to scale *any* shape, including the strange blob below, so that its perimeter equals its area! Perimeter is a length,

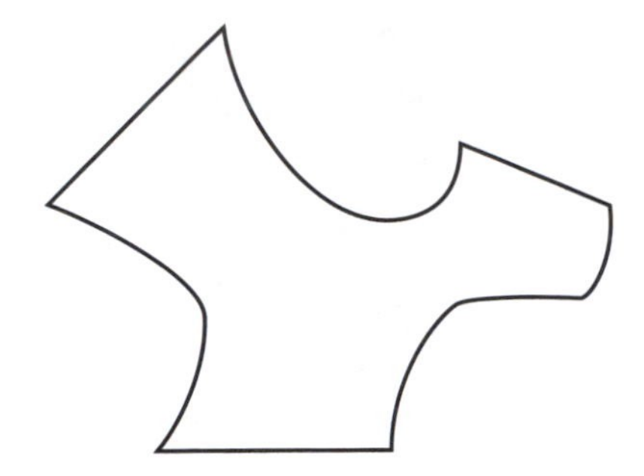

and area is a length squared. If we scale a picture by a factor k, then the perimeter changes by k whereas its area changes by the factor k^2. As any line $y = c_1 k$ and quadratic $y = c_2 k^2$ must intersect, there must be a scale for which a given shape has equal perimeter and area.

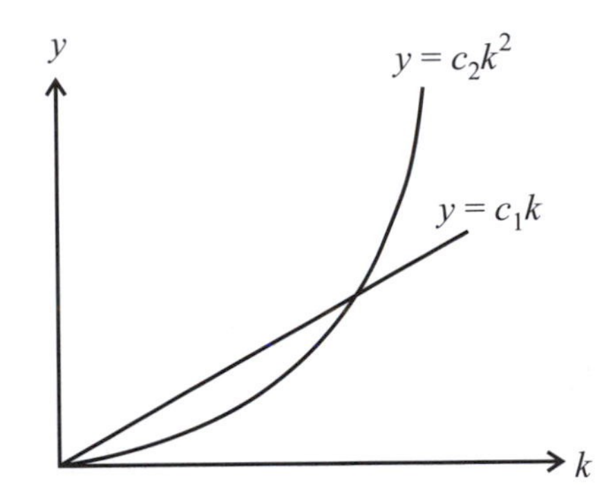

Challenge. With the use of a photocopier, can you produce a reduced copy of this very page with perimeter equal to area (measured in inches and square inches)?

It is not hard to see that an equilateral triangle scaled this way must have side length $4\sqrt{3}$. An $a \times b$ rectangle with this property must satisfy

$$a = 2 + \frac{4}{b-2}$$

and thus a 3×6 rectangle is the only such rectangle with integer side lengths. (The other solution, $a = 4$, $b = 4$, yields a square.) For a right triangle (a, b, c), solving for c in $\frac{1}{2}ab = a + b + c$ and substituting the result into $a^2 + b^2 = c^2$ yields

$$b = 4 + \frac{8}{a-4},$$

for which only $a = 5, 6, 8$, and 12 yield integer solutions. This gives just two integer-length right triangles with equal perimeter and area: the (5, 12, 13) triangle and the (6, 8, 10) triangle.

Taking It Further Answer. Unfortunately, there are no rectangular boxes with equal volume, surface area, and edge sum. To see this (if you are game!), first solve for c in the equation $abc = 4(a + b + c)$ and substitute into $4(a + b + c) = 2(ab + bc + ca)$. Regard the resultant equation (with denominator multiplied through) as a quadratic in b whose discriminant, as it turns out, is a quadratic in a that is always negative!

Hard Challenge. My students and I are wondering whether there exist any polygonal-faced solids with equal volume, surface area, and edge sum. Can you construct an example of such an object?

A Note on Fractal Dimensions

The shapes presented in this exercise have perimeters claiming to be areas. Their success, however, is but an artifice of scale. It is a surprise to learn that there exist examples of curves and perimeters that legitimately are more than mere "lengths." In 1904 the Swedish mathematician Helge von Koch was one of the first people to describe such a curve.

His construction begins with four straight line segments, equal in length, arranged to form a triangular hump. Replacing the middle third of each segment with a small triangular hump, one third the scale of the original hump, produces a secondary 12-edged figure. Replacing the middle third of each of these 12 new edges by another triangular hump, one third the previous scale, yields a new figure with new edges on which to perform the same operation. Repeating this process over and over again, ad infinitum, results in a curve-like construction possessing a certain degree of "fuzziness." This construct is today called the *Koch curve*.

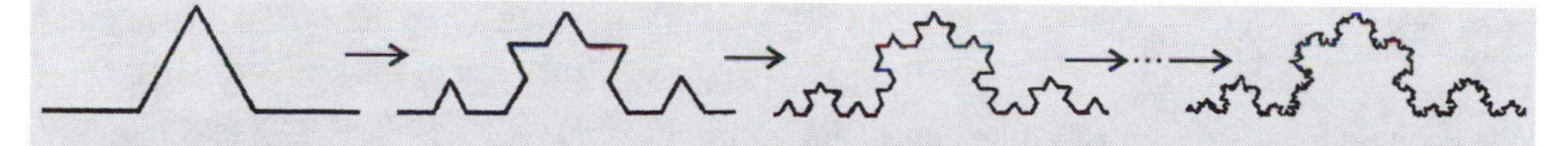

We think of a length as one-dimensional, meaning that if we scale a picture of a length, such as a perimeter, by a factor of k, the length, or "size," of the new picture is $k^1 = k$ times the size of the original.

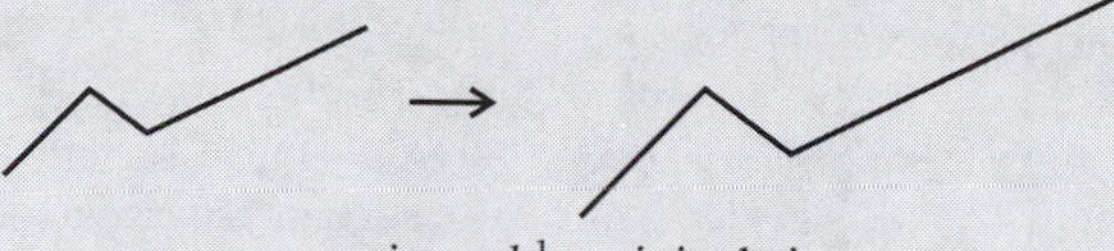

new size = $k^1 \cdot$ original size

An area is two-dimensional. Scaling a picture of an area by a factor k changes its "size" by a factor of k^2.

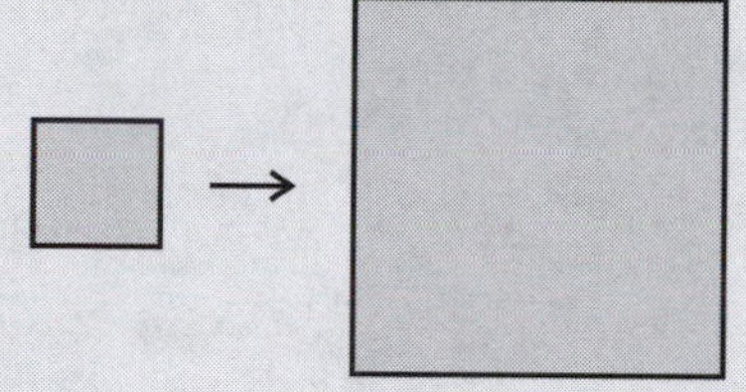

new size= $k^2 \cdot$ original size

Let d be the dimension of the Koch curve. As this curve consists of four scaled copies of itself, the size of the total curve is four times the size of each scaled-down version (each copy is scaled by a factor of one third). By the analysis above, the size of each scaled down version is k^d times the size of the whole original picture (with $k = 1/3$). This gives

$$\text{size of curve} = 4 \times \left(\frac{1}{3}\right)^d \cdot \text{size of curve}.$$

Consequently $1 = 4/3^d$ or $3^d = 4$. So $d \ln 3 = \ln 4$ or

$$d = \frac{\ln 4}{\ln 3} \approx 1.26.$$

Although not enough to be two-dimensional (an area), this curve is certainly more than one-dimensional! Placing three copies of this curve to form a "snowflake" gives an example of a perimeter that is legitimately more than just a length!

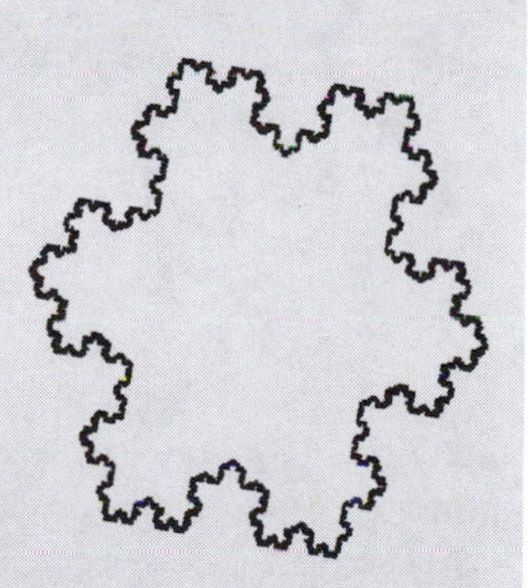

There are many examples of curves, shapes, and figures that lie between being lengths and areas. As an exercise, try calculating the dimension of the *Sierpinski triangle*. It is obtained from a solid equilateral triangle by removing the central equilateral triangle and repeatedly doing the same for any subtriangle consequently produced.

2.2. Weird Wheels

We essentially need to draw a circle but change our minds about what its center should be as we go along. We also need to make sure we return to the same center after every turn of 180°. Here's an easy way to do this using only a pencil, a straightedge and a piece of string.

Draw a collection of straight lines on a piece of paper (no two parallel), and choose one point on one line. Begin drawing a circular arc to the right from this point, with its center being the intersection of this line and the first line to the right of the point. When the arc hits that second line, continue with a new arc whose center is the point of intersection of this second line and the first line to the right of it. Continue all the way around the diagram. Notice that you return to the same center point every 180°!

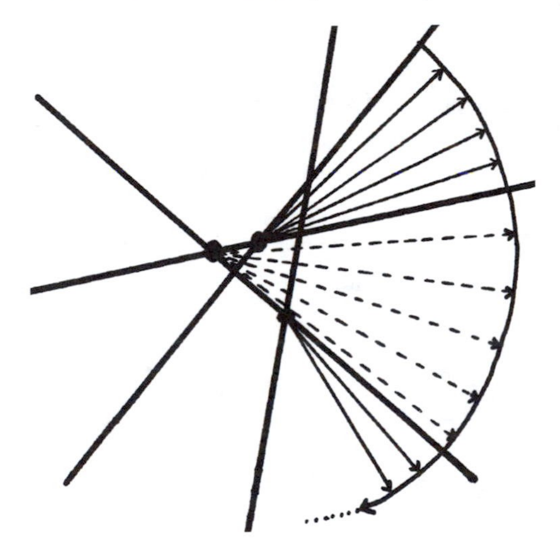

The height of this wheel equals the sum of the two radii of opposite sectors in this construction. Moving around the wheel one can easily see that this sum remains constant.

Question. What would it be like to ride in a car with wheels this shape?

Challenge. There are three-dimensional solids (other than spheres) of constant height. Can you envision one? (Have a look at Martin Gardner's article [Gard5].)

Taking It Further. Have two or more people construct different figures of constant height using the same height h. With a piece of string compare the perimeters of these shapes. What do you notice?

It turns out that all figures of constant height h have the same perimeter, namely πh, the perimeter of a circle of height (diameter) h. For an elementary proof of this, see Ross Honsberger's wonderful essay 17 in [Hons1].

2.3 Square Pegs and Not-so-Round Holes

Alas, I know of no simple ruler and string construction that produces the type of shape we seek. (Did you find one?) But new shapes with the desired property do exist and can be constructed with the aid of a computer.

As with problem 2.2, work with the parametric equations of a circle, but "wobble" the center this time in such a way that it returns to the same location after every 90° degrees of rotation. For example, center $(0, \frac{1}{20}\cos(4\theta))$ will do, leading to the equations

$$(0+\cos\theta, \tfrac{1}{20}\cos(4\theta)+\sin\theta) \quad 0 \le \theta \le 360°$$

and the shape shown here.

If you photocopy and enlarge this shape and the square within it, you can verify that this shape indeed has the desired property: All four corners of the square just touch the curve no matter the orientation of the square. (This shape also has the constant height property.)

Comment. Be careful with the choice of coefficients in the equations of this construction. The fraction $\frac{1}{20}$ was chosen so as to pro-

Josh Davis examines a square peg in round and not-so-round holes.

duce a shape that is convex. A coefficient of 1, for example, yields a non-convex hole.

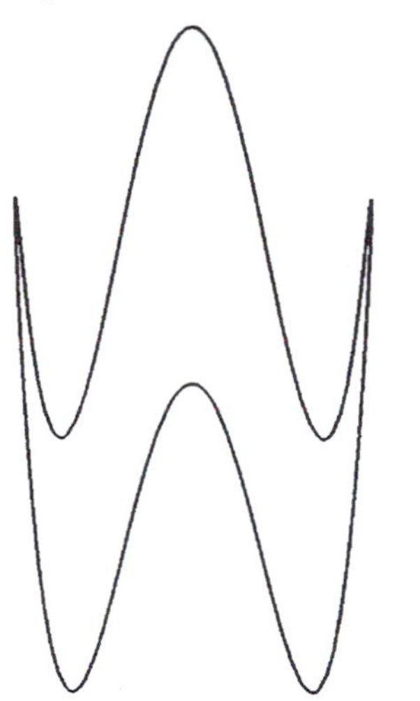

Challenge. Are there *non*-circular holes that will accommodate triangular pegs nicely? Equilateral triangles? *Non*-equilateral triangles?

Hard Challenge. A spherical hollow accommodates a cubical block in such a way that all eight corners of the block just touch the walls of the hollow, no matter the orientation of the block. Are there other shaped hollows with this property?

To conclude this discussion let's reverse the problem: Which shapes, other than circles, fit snugly into square holes so as to just touch all four edges of the square, irrespective of their orientation? The answer is precisely the shapes of Section 2.2—the figures of constant height! Can you see why?

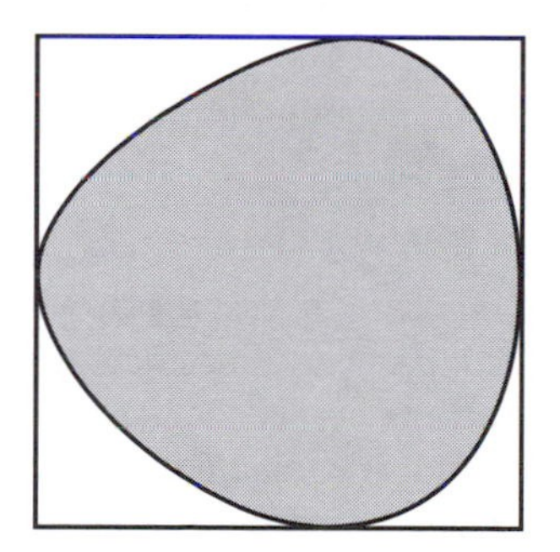

Acknowledgments and Further Reading

There are many elementary texts on fractal geometry. For example, [Deva2] is particularly fun and accessible; so too is [Barn], taking the work a little farther. For more properties of constant height figures, in addition to those already cited, see chapter 25 of Hans Rademacher and Otto Toeplitz's classic book [Rade]. The marvellous Square Peg problem appears in J. Konhauser, D. Velleman, and S. Wagon's wonderful book [Konh]. There they also do careful analysis to show, with a suitable choice of coefficients, that the computer- drawn curves really are convex.

3 Counting the Odds ... and Evens

3.1 A Coin Trick

Every flip changes the parity of the number of heads showing. Thus if an even number of heads was showing initially and an even number of flips was announced, an even number of heads should be showing in the end. If not, one head is covered by John's hand. The cases involving an odd number of flips or an odd number of heads initially present are analyzed similarly.

3.2 Let's Shake Hands

Following the notation of the hint, recall that N is odd. The sum

$$n_1 + n_2 + \cdots + n_N$$

is the total number of handshakes in which people are involved. As every single handshake represents two counts in this sum (two people each count "one" for the same handshake) this sum must be an even number—in fact, twice the total number of handshakes that occur. It is impossible then for all the numbers n_i to be odd, for a sum of an odd number of odd numbers is odd, not even. It is impossible then for an odd number of people to shake hands an odd number of times! (How did your group react to this predicament?)

One can extend this argument to show that if an arbitrary number of people shake hands an arbitrary number of times, then the total number of people involved in an odd number of handshakes must be even. (Try it! Verify by experiment this is always the case.)

3.3 Forty-Five Cups

Suppose at some stage of play you are left with r upright cups. If in a maneuver you decide to turn x of these upright cups over, $0 \leq x \leq 6$ (and hence $6 - x$ of the remaining cups), the number of upright cups then becomes

$$r \rightarrow (r - x) + (6 - x) = r + (6 - 2x)$$

The value of r can thus change only by an even amount. Initially $r = 45$, so it is impossible to attain $r = 0$. Thus the puzzle is unsolvable.

One can also analyze the puzzle this way: To obtain a state in which every cup is upside down, each cup must be turned an odd number of times. With 45 cups this means an odd number of inversions must occur in total. But this will never be the case as one is required to invert cups six at a time.

Comment. This argument also proves that it is impossible to invert any odd number of cups by turning over any even number of them at a time.

Challenge 1. Show that it is possible to invert 40 cups, turning over six at a time. Show that it is possible to invert 41 cups, turning over five at a time.

Challenge 2. Is it possible to invert 41 cups, turning over 37 at a time? Develop a general theory as to when it is possible to turn a cups over, b at a time.

3.4 More Plastic Cups

Those cups in positions corresponding to the perfect squares are upright, all the others are left upside down. The divisors of any number n come in pairs, namely d and n/d if d is a divisor, so every number has an even number of divisors *unless* for some divisor d we have $d = n/d$. In this case $n = d^2$ is a perfect square and has an odd number of factors.

Acknowledgments and Further Reading

Parity problems like these abound in popular mathematics books and are often used in magic tricks. For more puzzles of this type, see [Gard1], [Gard2], [Gard7], [Gard14], and chapter 29 of this book. The handshake problem represents a well-known result in graph theory: *The number of odd degree vertices in any graph must be even.* To learn more about the elements of this subject see Gary Chartrand's wonderful book [Char]. The Forty-Five Cups problem is a classic puzzler. A 1001-cups version of it appears in [Chan]. Problem 3.4 appears in many books under different guises—a warden turns the keys of a row of prison cells, or a schoolboy turns the keys of a row of student lockers, for example. This latter version appears in Ravi Vakil's inspiring book [Vaki]. See also [Gard16], chapter 6.

4 Dicing, Slicing, and Avoiding the Bad Bits

4.1 Efficient Tofu Cutting

Consider the innermost cube. It has six faces, each of which must be cut. As no two can be cut simultaneously with a single slice, a minimum of six slices is needed to perform the task.

Taking It Further Hint After some experimentation you have probably discovered that, with stacking, it is possible to dice a $4 \times 4 \times 4$ cube in just six slices and a $5 \times 5 \times 5$ cube in nine, and you probably suspect these numbers are minimal. Keep experimenting with larger-sized cubes and see if you can develop a general formula for the minimal number of slices needed to dice an $n \times n \times n$ cube into n^3 smaller cubes. Draw a table of your results and see if a pattern emerges. The efficient tofu dicing tactics you develop in playing with this problem might lead you not only to the correct minimal number formula but also to a proof that your formula is correct. Keep playing! (All the answers, analysis, and derivations are revealed in section 4.2.)

4.2 Efficient Paper Slicing

In slicing tofu, paper, and string, you have probably deduced that slicing pieces in half, or as close to half as possible, at every stage yields the most efficient dicing, slicing, or cutting method. Thus an efficient dissection algorithm utilizes the divisibility by 2 of all the numbers involved. For example, after some experimentation, you might establish that three cuts are necessary to dice an $n \times n \times n$ cube with $n = 2$; six cuts for $n = 3$ or 4; nine for $n = 5, 6, 7,$ or 8; 12 for $n = 9, \ldots, 16$; and so on. The powers of 2 represent places of change.

The following Note on Minimality Formulae establishes that the minimal number of cuts required is given by the formula $3\lceil \log_2 n \rceil$. (Here, for a real number x, $\lceil x \rceil$ denotes the least integer greater than or equal to x.) Did you guess this?

A Note on Minimality Formulae for Slicing and Dicing

Warning. Familiarity with logarithms and proofs by induction is required!

We begin by establishing some notation. For positive integers a, b, and c, let

$m(a) =$ minimal number of cuts needed to divide a string a units long into a unit segments (with stacking allowed).

$m(a, b) =$ minimal number of straight line cuts needed to slice an $a \times b$ rectangle into ab unit squares (via stacking).

$m(a, b, c) =$ minimal number of planar cuts needed to dice an $a \times b \times c$ parallelepiped into abc unit cubes (via stacking).

Lemma 1. $m(a) = \lceil \log_2 a \rceil$.

Proof. We will prove this via induction on a, the result clearly being true for a string of length 1. Assume the result is true for strings of length $< a$, and consider the subdivision of a string of length a.

We need to make an initial cut. This divides the string into two pieces of length a_1 and a_2, with $a_1 \geq a_2$ say. By the induction hypothesis we know we can cut the piece of length a_1 in a minimum of $m(a_1) = \lceil \log_2 a_1 \rceil$ cuts. Since $m(-)$ is clearly a non-decreasing function (longer strings take at least as many cuts as shorter strings), we can cut the other piece in probably less than, but certainly no more than, $m(a_1)$ cuts. Thus, imagining we are cutting the piece of length a_2 along with the piece of length a_1 (via stacking), we need only focus our attention on this larger piece of length a_1.

In order to find the minimal number of cuts required to cut the original string of length a, we need to choose a value of a_1 that yields the minimal value for $m(a_1)$. We want to choose a_1 as small as possible then (and still at least half the length of the string). We must choose

$$a_1 = \begin{cases} \dfrac{a}{2} & \text{if } a \text{ is even,} \\ \dfrac{a+1}{2} & \text{if } a \text{ is odd.} \end{cases}$$

Now logarithms work as follows. If $2^{n-1} < a \leq 2^n$ for some integer n, then $\lceil \log_2 a \rceil = n$. Thus if a is even, $2^{n-2} < a/2 \leq 2^{n-1}$ and $\lceil \log_2 a/2 \rceil = n - 1$. If a is odd, $2^{n-1} < a + 1 < 2^n$ and $\lceil \log_2 (a+1)/2 \rceil = n - 1$. In either case our choice of a_1 yields $m(a_1) = \lceil \log_2 a_1 \rceil = \lceil \log_2 a \rceil - 1$. Thus

$$m(a) = m(a_1) + 1 = \lceil \log_2 a \rceil.$$

This completes the proof by induction. ■

Lemma 2. $m(a, b) = m(a) + m(b)$.

Proof. Use induction on the sum $k = a + b$. If $k = 2$ the result is clearly true.

Assume the result is true for all values $< k$ and consider the case of an $a \times b$ rectangle with $a + b = k$. We commence with an initial cut. Assume this results in two pieces of sizes $a_1 \times b$ and $a_2 \times b$ with $a_1 \geq a_2$. Because of stacking, we need only focus on the $a_1 \times b$ piece. By the induction hypothesis this can be cut in a minimum of $m(a_1, b) = m(a_1) + m(b)$ slices. We thus need to find a choice of a_1 that minimizes this formula. Since $m(-)$ is a non-decreasing function, we must choose $a_1 = a/2$ or $(a + 1)/2$ depending on whether a is even or odd. Thus

$$\begin{aligned} m(a,b) &= m(a_1,b)+1 \\ &= m(a_1)+m(b)+1 \\ &= \lceil \log_2 a \rceil - 1 + m(b) + 1 \\ &= m(a)+m(b). \end{aligned}$$

This completes the proof by induction. ■

Lemma 3. $m(a, b, c) = m(a) + m(b) + m(c)$.

Proof. Analogous to the proof of Lemma 2. ■

Thus the minimal number of slices required to slice an $n \times n \times n$ cube of tofu is $m(n,n,n) = 3m(n) = 3\lceil \log_2 n \rceil$.

Surprise. We have just proven that it is possible to dice a cube of tofu into 1,000,000 little cubes in just 21 slices! (And no fewer!)

Challenge 1. I have drawn k straight lines through a common point on a piece of paper. What is the minimal number of cuts (via stacking) required to do the indicated slicing? What about other patterns of lines? Can you obtain minimal cutting results for other configurations?

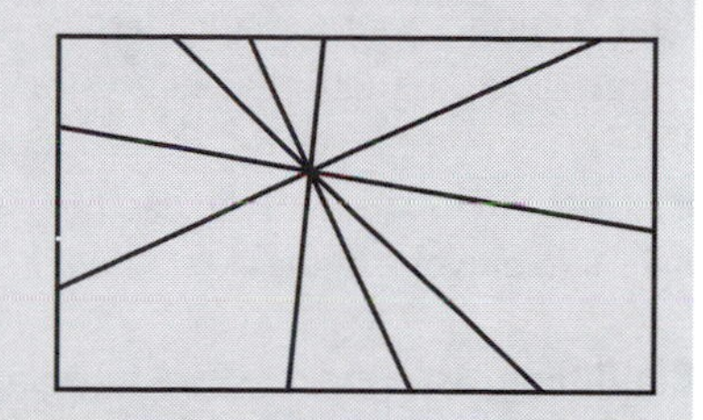

Challenge 2. A piece of paper in the shape of an equilateral triangle with a side length of n units is to be sliced into n^2 subtriangles. Via stacking, what is the minimal number of straight line cuts required to accomplish this feat?

As a partial solution you could certainly dissect the figure as though it were half an $n \times n$ rectangle problem, and then simultaneously slice each resulting "subsquare" diagonally in half with one additional slice. This requires a total of $m(n,n) + 1 = 2\lceil \log_2 n \rceil + 1$ cuts. Is this number minimal?

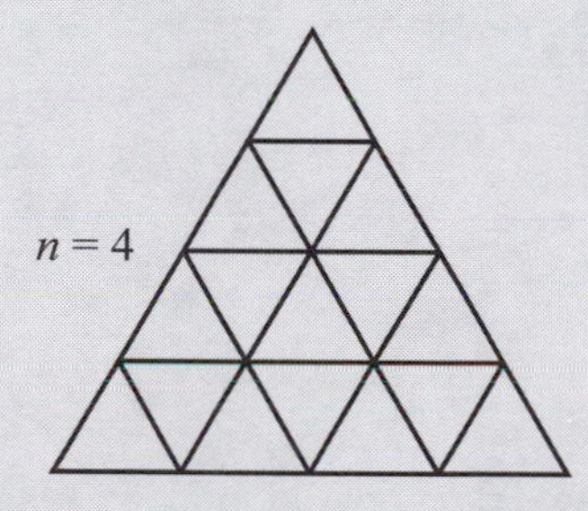

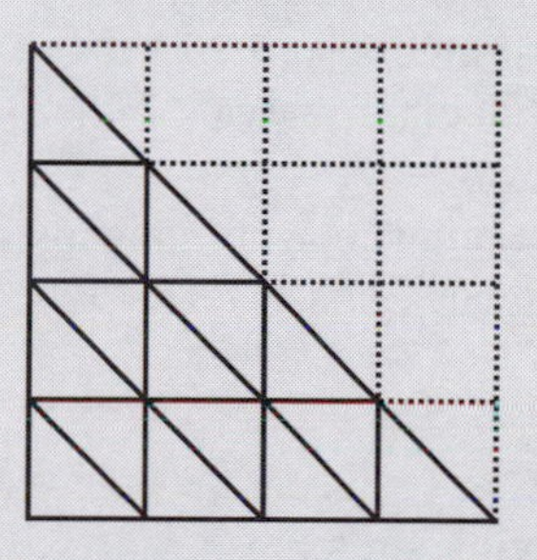

Taking It Further. Taking it up a dimension, here's a trick question: What is the minimal number of planar slices required to dice a regular tetrahedral piece of tofu into identically sized subtetrahedra?

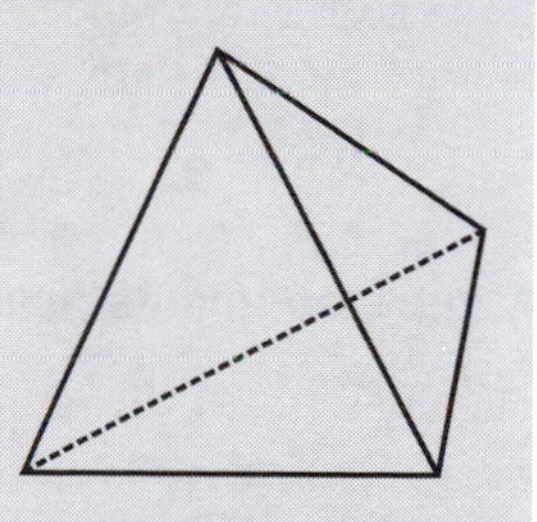

The answer is to give up! It is impossible to subdivide a tetrahedron into subtetrahedra in this way. (Try it!) For a proof of the impossibility of this task, see chapter 5 of Stan Wagon's fabulous (advanced) book [Wago].

Challenge 3. Imagine you were allowed to both fold and stack pieces of paper before cutting. What is the minimal number of cuts needed to slice a rectangular $a \times b$ piece of paper in this case?

Challenge 4. Consider the reverse to our cutting problem. If you are allowed k cuts, what is the maximal number of pieces of paper you can obtain with stacking? Without stacking?

Suppose in k cuts you obtained the maximal number of pieces possible without stacking. Could you always obtain the same outcome (exact same pieces) in fewer than k cuts if stacking were allowed?

4.3 Bad Chocolate (Impossible!)

Dan can always guarantee a win by handing James a square grid of chocolate containing the bad piece. James will always be forced to hand back a non-square shape, hence, never the single piece. Exactly the same strategy works for the second player in the second game. Notice that the location of the bad piece of chocolate is irrelevant in these puzzles.

Challenge 1. A 4×8 rectangular bar of chocolate contains *two* bad squares located at diagonally opposite corners. At every move a player is handed either one piece of chocolate containing two bad squares, or two pieces containing one bad square each. The player chooses a piece, breaks off a segment along a score line, puts the good chocolate aside, and hands the bad chocolate back over. As soon as someone is handed both bad squares of chocolate as two single pieces, the game is over.

Is there an optimal playing strategy for this game? What about a game with a 4×9 rectangular block?

Challenge 2. Can you generalize these games and their playing strategies to three dimensions with cubic blocks of chocolate? What about flat equilateral triangular bars of chocolate?

Acknowledgments and Further Reading

The cube (tofu) slicing problem is a teaser I was told as a school student. I mention it in my article [Tant3] but haven't been able to track down its origin. Other slicing problems and games appear in the amazing interactive website supervised by Alexander Bogomolny [Bogo] (see "Breaking Chocolate Bars") and in E. Berlekamp, J. H. Conway, and R. K. Guy's fabulous text [Berl1]. The classic book on slicing with planes is [Orli].

My thanks to Dr. Eric McDowell for his input on cube slicing.

5 "Impossible Paper Tricks

5.1 A Big Hole

After completing the hint, insert a "zipper" of cuts as shown. This produces a hole big enough to walk through!

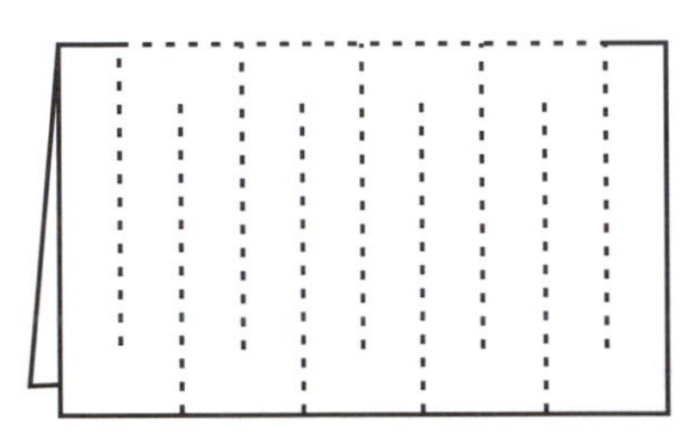

Kelly Ogden walks through an index card.

5.2 A Mysterious Flap

Continuing from the hint, a simple 180° rotation of the bottom half of paper produces the mysterious flap!

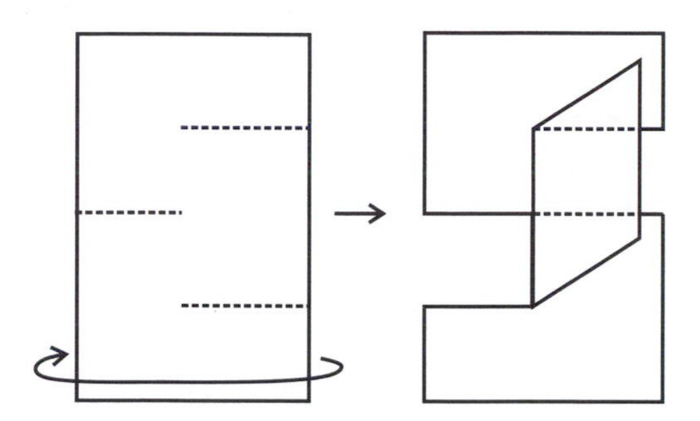

Advertise your next public event with signs made this way. I guarantee they will garner plenty of attention and invoke much conversation. Plus, they have the advantage of being conspicuous from great distances along corridors! To make signs like these, photocopy the text and decoration on *two sides* of the paper as shown below before cutting and twisting.

5.3 Bizarre Braids

It is not possible. Here's an intuitive argument why not. Let's call the crossover of two adjacent strands *positive* if the left strand crosses over the right, *negative* otherwise. Let the number of positive crossovers minus the number of negative crossovers be alled the *index* of a braid. Thus the four-braid we are trying to create has index −4. To obviate the issue of twists in strands, let's assume we are working with four thin strings attached at both ends to small squares of cardboard . Certainly if it is impossible to braid these four strings, it is impossible to braid four thick strands.

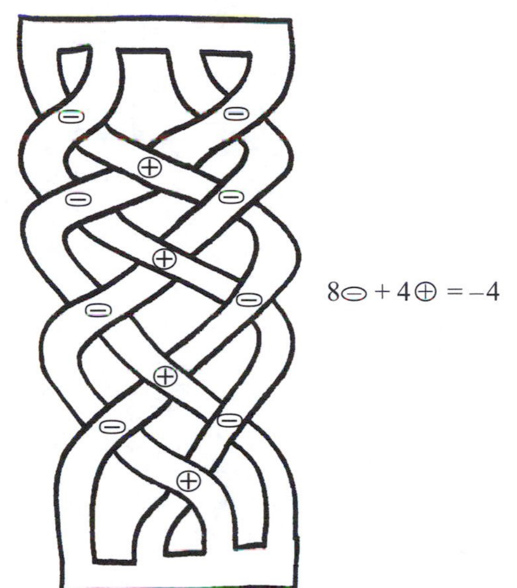

There are a number of basic maneuvers we could perform on this system. We could rotate

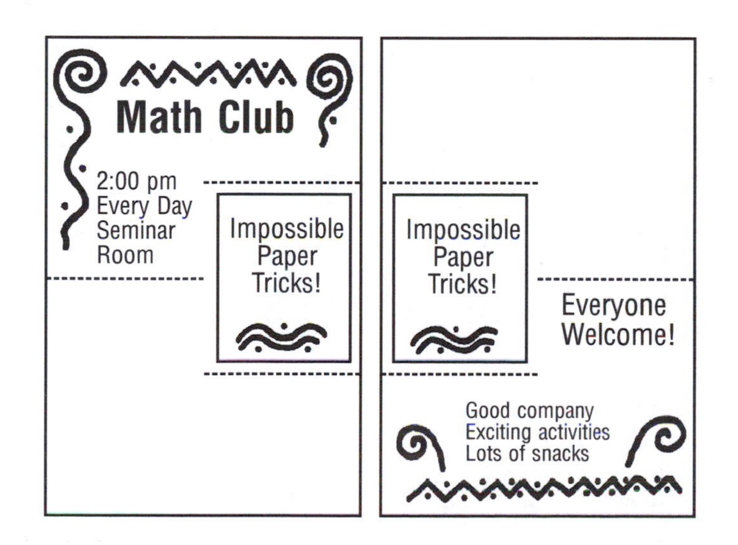

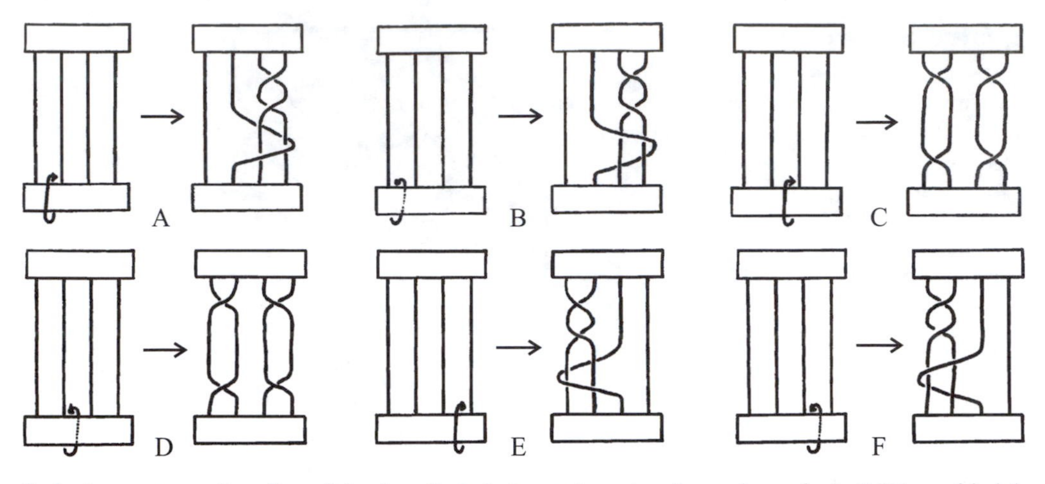

the bottom square of cardboard, back or forth, between two strands, as shown in A–F. We could pick one strand and rotate it, back and forth, around the bottom square of cardboard, as G–J illustrate.

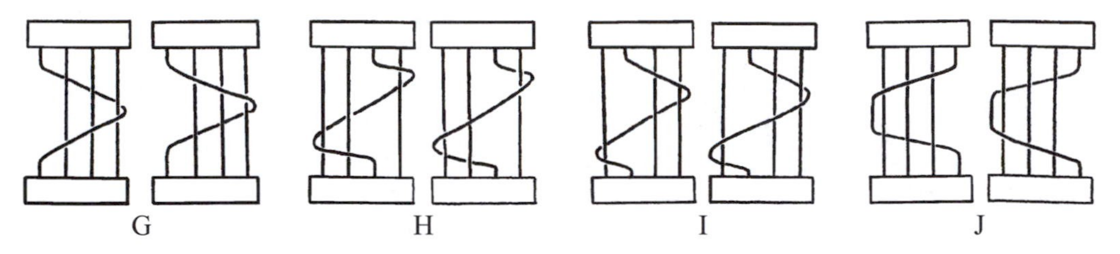

We could rotate the bottom square about a vertical axis, back or forth, as shown in K.

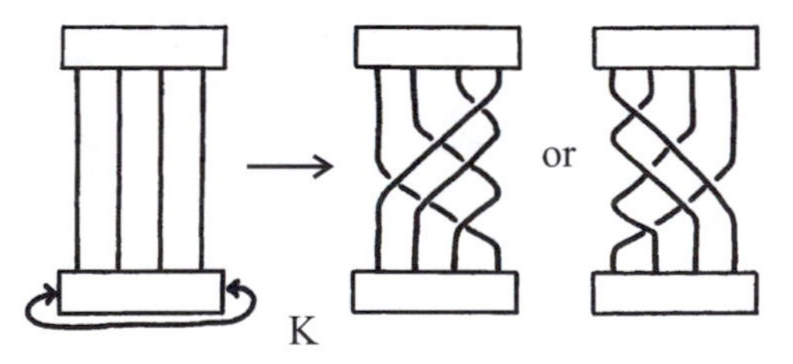

Or we could perform some fundamental maneuvers within a given diagram, as examples L–N suggest.

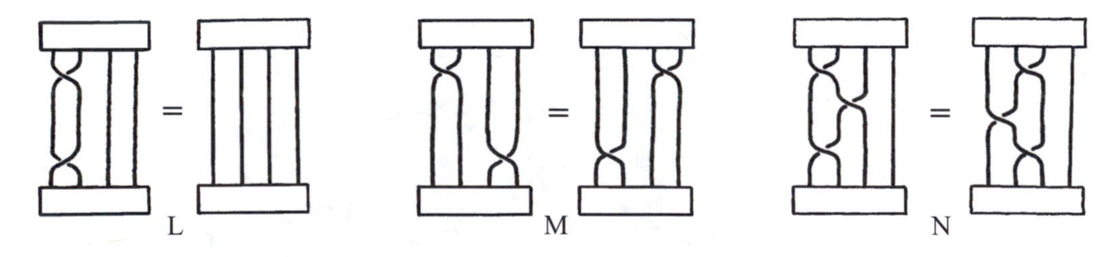

Each maneuver (except for the last three) introduces new crossovers into any given braid diagram and changes its index by $+6$, -6 or 0. We started with a diagram of four untangled strings, index 0, and hoped to achieve one of index -4. As every maneuver changes a diagram's index by a multiple of six, the task is impossible.

Comment 1. This argument can be made rigorous with the aid of Emil Artin's theory of braids. See [Arti1], [Arti2], and especially [Newm].

Comment 2. Tie three strings to the back of a chair and braid them (with free ends) in some complicated manner. (A braid is formed as a sequence of twists of two adjacent strands.) Just make sure when you are done that the end of the leftmost string ends up at the leftmost position, the end of the middle string in the middle, and the end of the rightmost string at the right.

Now tape these three ends to a small piece of cardboard making sure to preserve their order. Is it possible to untangle your braid with the three ends now fixed in place? It turns out the answer is always yes! No matter how complicated a braid you make with three free strands, exactly the same feat can be accomplished with the three ends tied together! Try it! (See [Shep].)

Challenge 1. Is it possible to make the following braid with no free ends? (Again, see [Shep].)

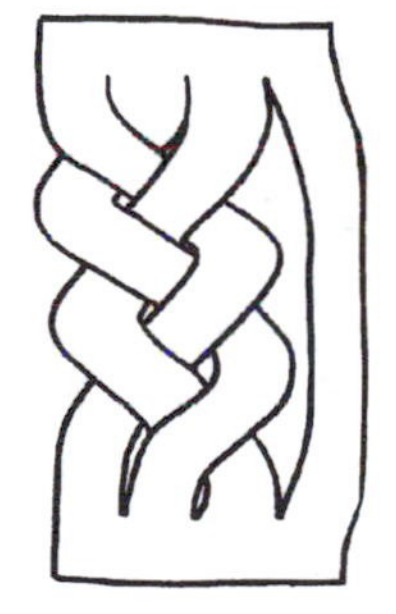

Challenge 2. Three strings are attached to a teacup and to three points about a room. The teacup is given a full turn of 360°. Without moving the teacup, prove it is impossible to maneuver the strings around the cup and untangle them . (See Section 12.1 and [Newm].)

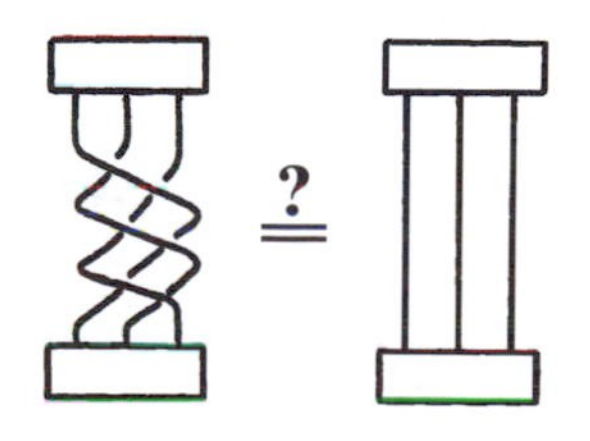

5.4 Linked Unlinked Rings

By replacing one ring in a pair of linked rings by a double band of rings we obtain the Borromean rings.

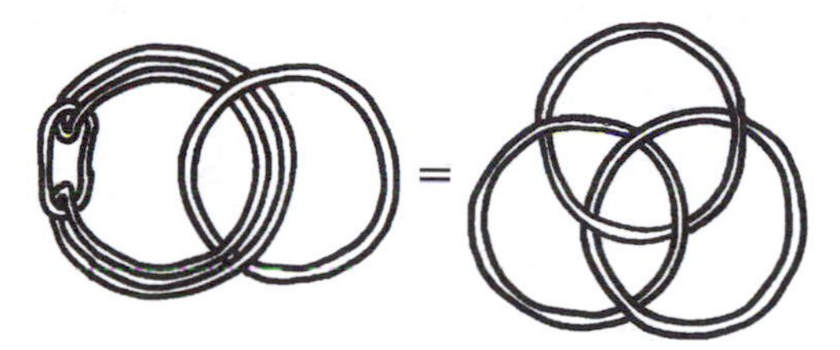

A foursome of linked unlinked rings is obtained from the construction shown.

It is also possible to construct a five-some and a six-some of rings:

Challenge. Can you construct seven or more linked unlinked rings?

Acknowledgments and Further Reading

I first learned of the "mysterious flap" from an advertisement for a social function with the geology department at MIT. Its appeal clearly spreads across all disciplines! Martin Gardner writes about the flap (he calls it a hypercard) in his essay on minimal sculpture, Chapter 8 in [Gard20]. Turner Bohlen, age 8, recently created *two* mysterious flaps in a single poster by making four horizontal cuts to the center line of a piece of paper, and applying two 180° twists. Other intriguing designs are possible. Martin Gardner also writes about the bizarre braid in [Gard6], Chapter 4, and in [Gard12]. It is a trick familiar to all those who work in leather craft. (Look closely at leather belts next time you go shopping!) For an excellent introduction to the group theory of braids see [Gard2], chapter 2. To learn more about which braids can and cannot be plaited see A. H. Shepperd's article [Shep]. For a different type of braiding problem (weaving) see [Farr]. A discussion on the Borromean rings appears in [Gard3].

6 Tiling Challenges

6.1 Checkerboard Tiling I

The excision of a black cell always leaves a tilable configuration. To see this, draw a path from the top left cell to the bottom right cell as shown.

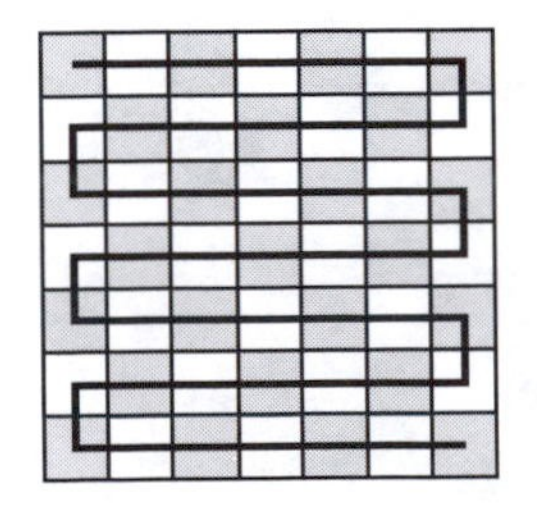

The removal of a black cell breaks this path into two segments of even length. Laying dominoes along these two trails then gives a tiling of the entire configuration!

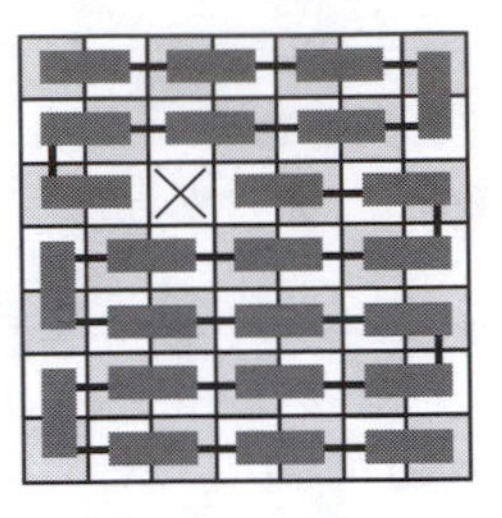

6.2 Checkerboard Tiling II

Taking It Further Answer. Yes! Again consider a path winding its way through the 8×8 array as shown. Removing any two cells of opposite color breaks this path into two segments of even length (or one piece of even length if the two cells happen to be adjacent along this path). Simply laying dominoes along these two trail segments produces a tiling of the entire surviving configuration!

Hard Challenge. Suppose two white cells and two black cells are excised from an 8×8 checkerboard. Is the surviving configuration guaranteed to be tilable? Just how many, or in what way, can cells be removed from an 8×8 board to guarantee tilability?

Tim Cavanaugh of St. Mary's College of Maryland (Class of '99) wrote a fantastic interactive program that quickly tests the tilabilty of square lattice regions [Cava]. You can design lattice regions and quickly add and remove cells in any way you like. You might like to play with his program to help explore this question.

6.3 Checkerboard Tiling III

Notice that in each coloring scheme each 3×1 tile must cover precisely one cell of each color,

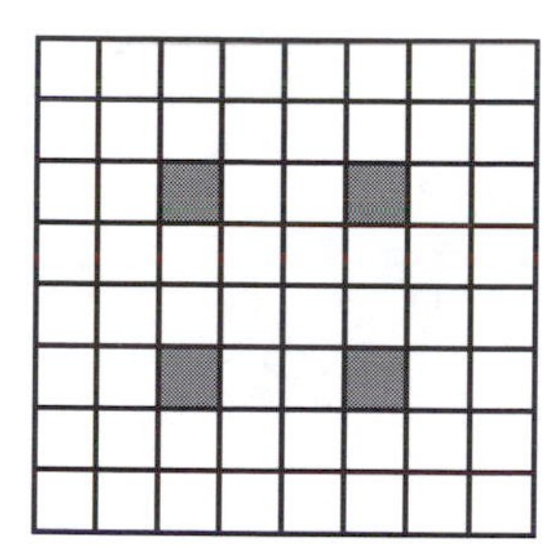

no matter where it is placed. As there are 21 white cells, 21 black cells, and 22 grey cells in the first diagram, the single tile must be placed on a grey cell. By the same reasoning it must be placed in the location of a grey cell on the second diagram too. This leaves only four possible locations of the single tile—all of which are equivalent given the rotational symmetry of the square. Once the single tile is placed at one of these locations, tiling the remainder of the diagram is not difficult.

Challenge. Let n be any integer not divisible by three. Is it always possible to tile an $n \times n$ square grid with a number of 3×1 tiles (in fact, with $(n^2 - 1)/3$ of them) and a single 1×1 tile?

6.4 Checkerboard Tiling IV

Such a tiling is indeed impossible. As suggested by the hint, consider just one line and the number of tiles that cross it in any given tiling. First, note that it cannot be the case that the number of tiles crossing this line is odd, for this would leave an upper (and lower) region with an odd number of cells to be tiled by dominoes, which is impossible. Thus the number of tiles crossing it is even. If this even number is zero, then it is a separating line. So to avoid separating lines, each line must be crossed by at least two tiles. As there are five horizontal and five vertical lines, we need at least 20 tiles to accomplish this feat. But only 18 tiles are available to us, so it cannot be done!

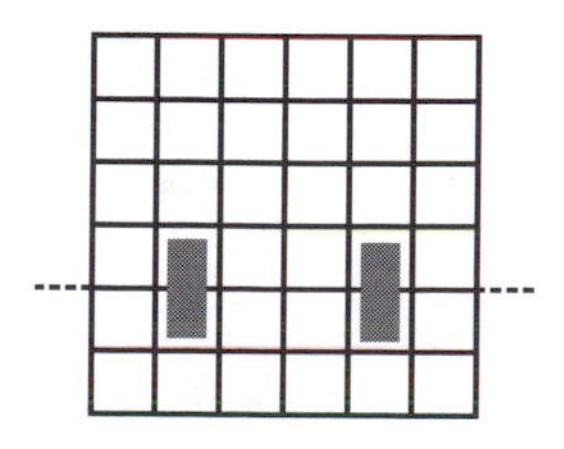

Challenge 1. Is it possible to tile an 8×8 array with no lines of separation?

Challenge 2. Lines of separation also occur when tiling a 4×4 array and a 4×12 rectangular array. Determine a complete list of array sizes, $a \times b$ (with at least one of a and b even), for which lines of separation are guaranteed.

Acknowledgments and Further Reading

The issue of domino tiling on square lattices is an old one. Given the checkerboard coloring scheme, any region tilable with dominoes must contain an even number of black and white cells. But the converse need not be true. The following region, for example, satisfies this criterion but is not tilable. Bill Thurston [Thur] characterized those regions that can be tiled by dominoes; he also discusses tilings on other lattices. Nonetheless, domino tiling problems are still a source of much amusement and challenge [Gard2], [Gard3], [Gard10], [Golo], [Sing], [Vaki]. The Separating Lines puzzle appears in [Hons2], chapter 6.

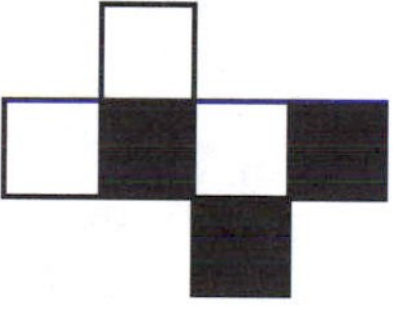

7 Things That Won't Fall Down

7.1 Wildly Wobbly

Six-faced polyhedral figures with precisely six, five, four, three and two stable faces certainly do exist. See the examples on the next page. My students and I haven't been able to con-

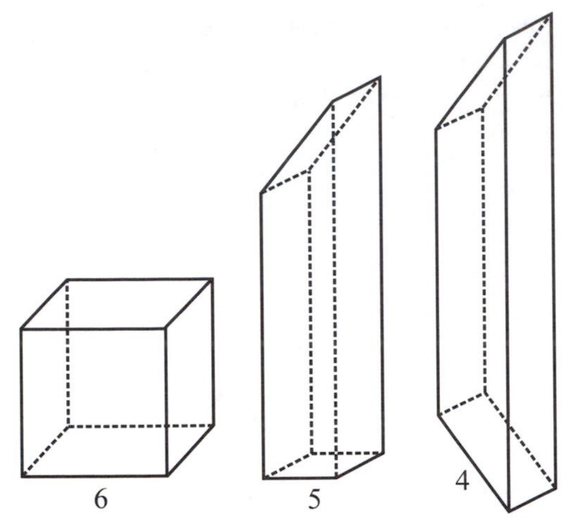

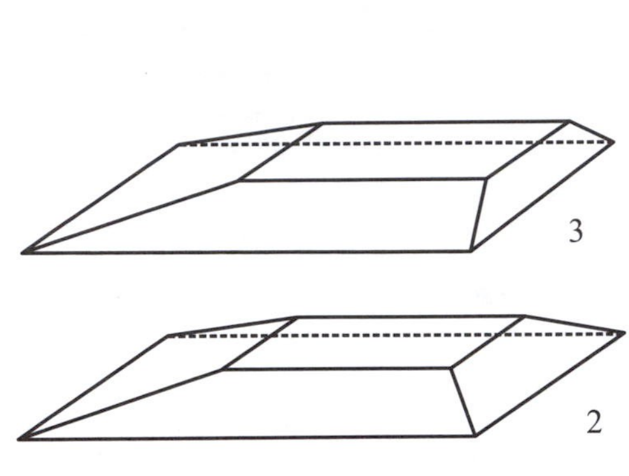

struct a six-faced model with only one stable face and, as far as I am aware, neither has anyone else on this planet! (Did you succeed?) *Unistable polyhedra* do exist; the first was discovered in 1968 by Richard K. Guy [Crof], [Conw1]. But no one has yet been able to build one with fewer than 19 faces.

My students have not been able to construct a *uni-stable wheel* either, and for good reason! Unlike their three dimensional counterparts, these objects cannot exist. The "Note on Polygonal Wheels" explains why.

A Note on Polygonal Wheels

Here we shall prove that every convex polygonal wheel made of uniformly dense material must possess at least two stable edges. (*Warning.* This proof is tricky and uses elements of calculus.)

A wheel will "topple over" from a given edge if its center of mass extends beyond the length of the edge. Suppose that, contrary to our claim, a wheel with just one stable edge exists. Let C denote the position of the center of mass of this wheel. Measuring angles from a horizontal ray positioned at C, let $r(\theta)$ denote the length of the "spoke" from C to the boundary of the wheel at angle θ. This represents a continuous function of θ, $0 \le \theta \le 2\pi$, with $r(0) = r(2\pi)$.

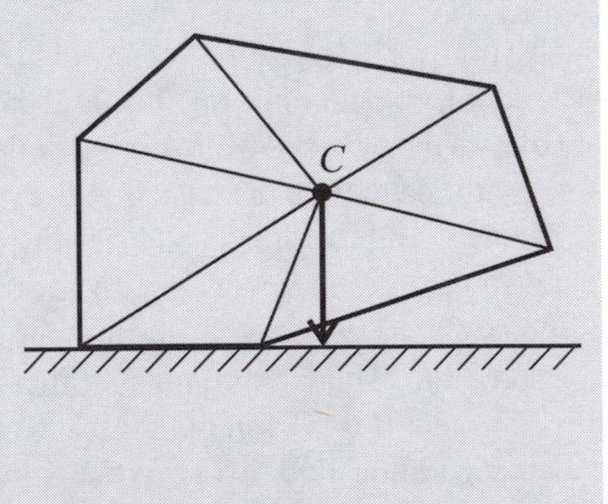

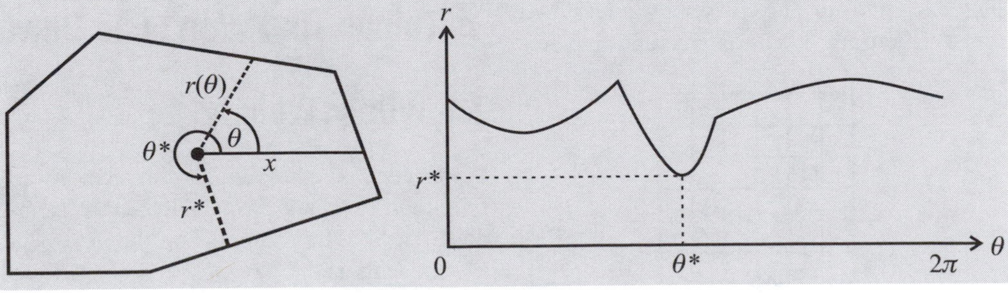

Every graph of a continuous function on such an interval possesses local maxima and minima. At a local minimum, $\theta = \theta^*$, $r(\theta^*)$ adopts a value r^* smaller than (or perhaps equal to) function values at nearby points. As the edges of the wheel are made of straight line segments and the wheel is convex, the spoke at angle θ^* meets an edge tangent to the circle of radius r^*. Thus the center of mass sits directly above this edge of the wheel and this edge is stable. We have assumed there is only one stable face, thus this function can have only one local minimum, and this unique local minimum is an absolute minimum.

Local maxima of the function $r(\theta)$ occur only at the vertices of the wheel, and thus at isolated points. By the Extreme Value Theorem, between any two local maxima there must be a minimum. Considering the "wraparound effect" of the domain of our function (recall that $r(0)$ and $r(2\pi)$ have the same value), if this function had two local maxima it would also have two local minima. This is not the case. Thus our radial function also possesses just one local maximum—the absolute maximum.

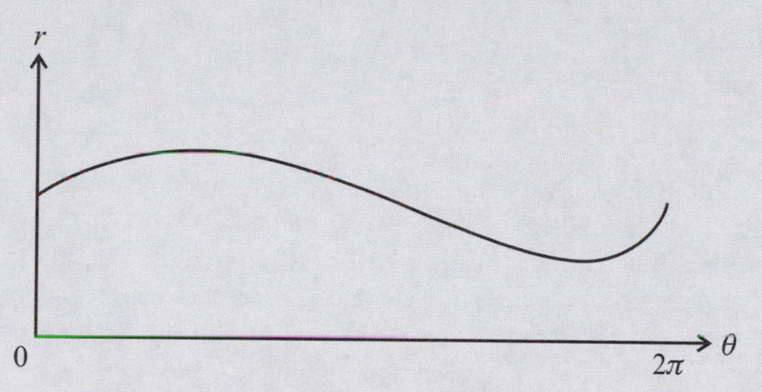

These two extrema are no more than π radians apart (consider the wraparound effect again). Assume, rotating the wheel if necessary, the maximum occurs at angle $\theta = 0$ and the minimum at angle $\theta = a$, with $0 < a \leq \pi$. (One may have to interchange the role of the maximum and the minimum here. The remainder of the proof however would be analogous.) Let $f(\theta) = r(\theta) - r(\theta + \pi)$, with the convention that an angle greater than 2π is equivalent to an angle 2π radians less. Then $f(0) > 0$, since $\theta = 0$ is a strict absolute maximum, and $f(a) < 0$, since $\theta = a$ is a strict absolute minimum. Consequently, by the Intermediate Value Theorem, there is an angle θ' between 0 and a with $f(\theta') = 0$; that is, with

$$r(\theta') = r(\theta' + \pi).$$

Denote this common value r'. Note that the minimum occurs within the interval $(\theta', \theta' + \pi)$ and necessarily $r(\theta) < r'$ for all $\theta \in (\theta', \theta' + \pi)$. For θ outside this interval, $r(\theta) > r'$.

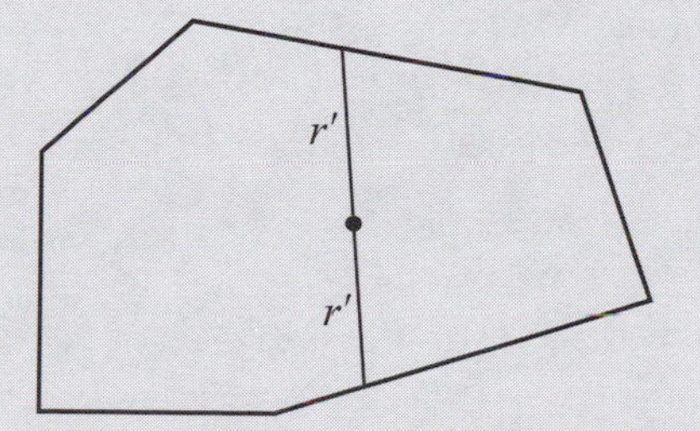

We're almost done. The final step is to analyze the areas of the two "halves" of the wheel divided by the line at $\theta = \theta'$. One "half" lies entirely within a semi-circle of radius r'. The other "half" extends beyond a semi-circle of radius r'. Since the density of the material is uniform, we see that the wheel would not balance on a ruler's edge placed along this line at angle $\theta = \theta'$. This line passes through C, the center of mass, and so should balance. We have reached a contradiction.

Thus there is no convex polygonal wheel with just one stable edge. ■

Challenge. Prove mathematically that every convex polygonal wheel, made of uniformly dense material, has at least one stable edge. (Thus there is no need to rely on the nonexistence of perpetual motion.)

Comment. Anyone willing to cheat, can construct polygonal wheels with single stable faces. Just force the center of mass to be at a certain position by inserting a lead weight. Here's an example.

lead weight

Connection to Uni-stable polyhedra. Although every convex wheel must possess at least two stable edges, convex polyhedra with only one stable face do exist. Given the analysis of the two-dimensional problem, this result seems very surprising. The key to understanding how such an object could possibly exist lies within our previous stipulation that the wheel me made of uniformly dense material. If the density of the material is not uniform, one can produce *uni-stable* wheels.

To build a uni-stable polyhedron try using the cross section of a cylinder. Angling the ends appropriately can displace the center of mass of the cross section until only one face of the cylinder remains stable. With luck, these ends might also be sufficiently angled so that they themselves do not represent stable faces.

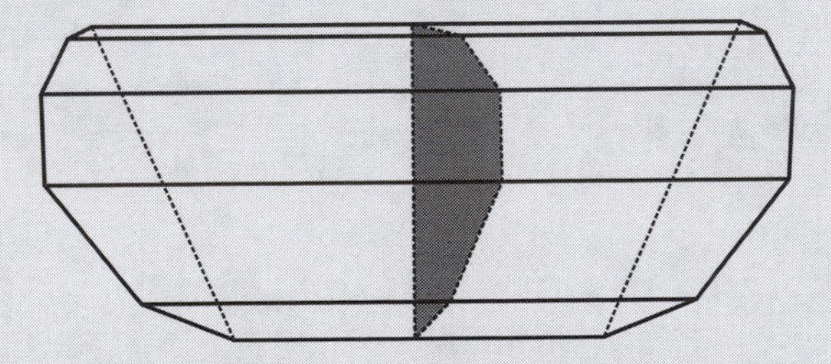

The record for the uni-stable polyhedron with the fewest faces is 19 [Conw1]. John H. Conway proved that no four-faced polyhedron (that is, no *tetrahedron*) can be uni-stable, but the minimal number of faces required is still an open problem.

Question. A *bi-stable polyhedron* is a polyhedron with just two stable faces. What is the least number of faces it can possess? Surprisingly the answer is just four! A. Heppes [Hepp] discovered the following tetrahedron with just two stable faces. Try building the model out of florist's foam!

41 20 17 26 4 24

7.2 A Troubling Mobile

Suppose $N-1$ wires have been balanced thus far and we wish to attach this system to the end of a top Nth wire. This is equivalent to a two-

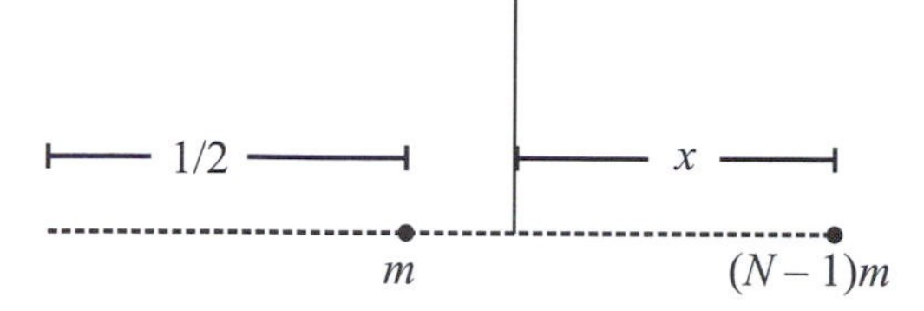

mass system. We wish to find the position x for which this system will balance. By Archimedes' Law of the Lever we must have: $(\frac{1}{2}-x)m = m(N-1)x$ from which we obtain $x = \frac{1}{2N}$. Thus a perfectly balanced mobile is constructed by measuring distances as indicated in the sketch.

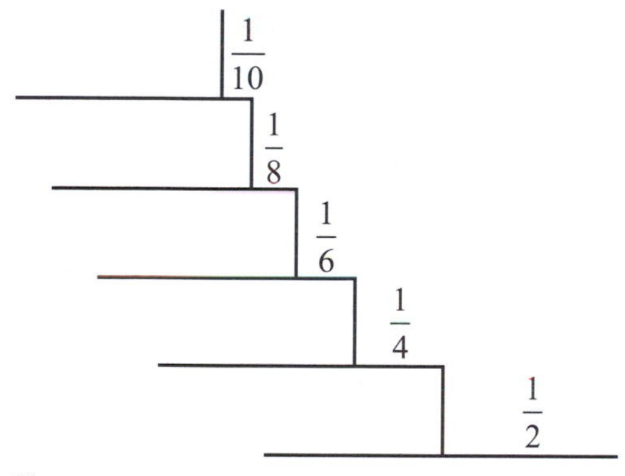

Since

$$\frac{1}{2}+\frac{1}{4}+\frac{1}{6}+\frac{1}{8}>1,$$

a mobile with five wires will theoretically extend beyond the length of the top wire. However, in practice, this is difficult to complete. It is hard to measure the distances precisely and usually more wires are needed. Students at St. Mary's College of Maryland managed to produce such a mobile with six wires but no one has yet managed five.

Taking It Further Answer. Since

$$\frac{1}{2}+\frac{1}{4}+\cdots+\frac{1}{60}>2,$$

a mobile of 31 wires would theoretically extend beyond more than two full wire lengths.

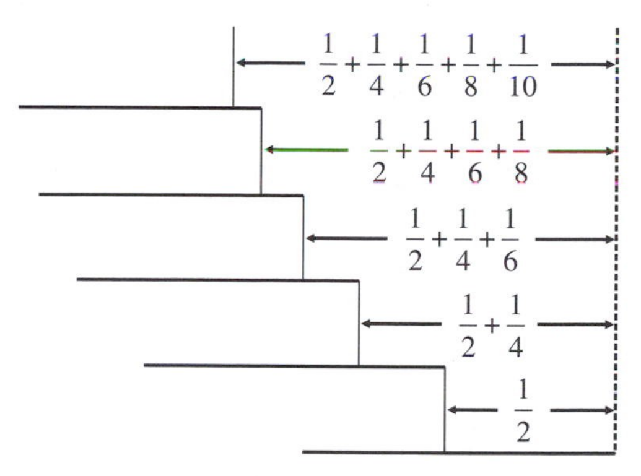

No one in our club has yet managed to build such a mobile. Can you?

Measuring from the right, as shown in the sketch above, the end points of this mobile lie at distances 0, $\frac{1}{2}$, $\frac{1}{2}+\frac{1}{4}$, $\frac{1}{2}+\frac{1}{4}+\frac{1}{6}$,.... In general, the nth wire lies at a distance

$$\frac{1}{2}\left(1+\frac{1}{2}+\frac{1}{3}+\cdots+\frac{1}{n-1}\right).$$

To get a handle on this sum, consider the integral $\int_1^n \frac{1}{x}\,dx$. To approximate this integral draw rectangles of unit width underneath the curve $y = \frac{1}{n}$ and sum the areas of these rectangles, as indicated in the graph. Thus

$$\frac{1}{2}+\frac{1}{3}+\frac{1}{4}+\cdots+\frac{1}{n}\approx\int_1^n \frac{1}{x}\,dx.$$

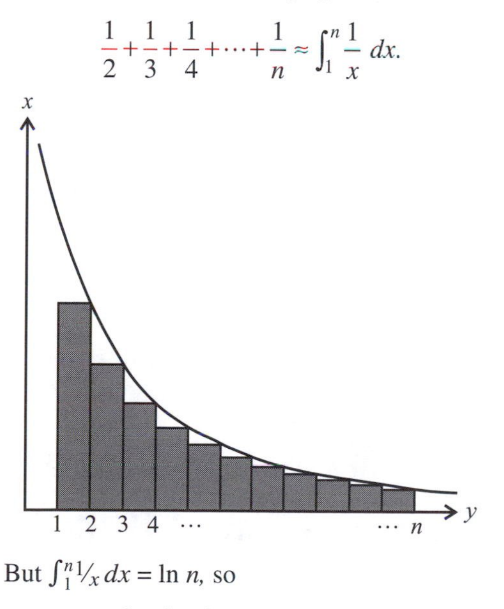

But $\int_1^n \frac{1}{x}\,dx = \ln n$, so

$$\frac{1}{2}+\frac{1}{3}+\frac{1}{4}+\cdots+\frac{1}{n}\approx \ln n.$$

Note that for large n there is an error of about

0.42278, which is $1-\gamma$ (*Euler's constant*). Thus the endpoints of the mobile's wires basically lie on a logarithmic curve!

Incidentally, it is well known that the series $1 + \frac{1}{2} + \frac{1}{3} + \frac{1}{4} + \cdots$ *diverges* (that is, summing sufficiently many initial terms of this series produces numbers as large as you like). This means that theoretically you can create mobiles that lean out for thousands of feet!

Challenge. How many wires are needed to construct a mobile that extends 100 full lengths?

7.3 A Troubling Tower

The construction illustrated here does the trick.

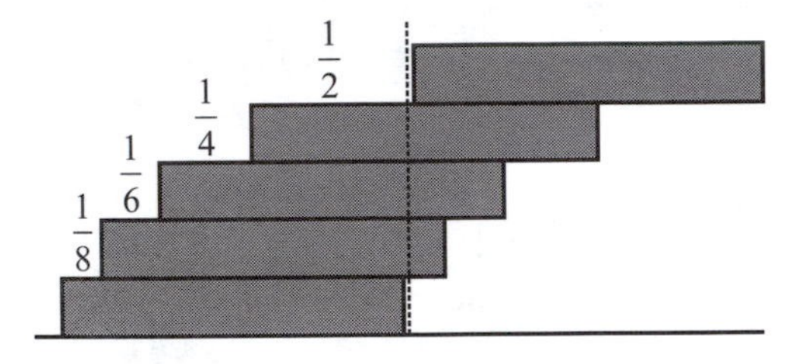

Acknowledgments and Further Reading

The stacking blocks problem is a classic demonstration in introductory physics courses [Eisn], [John]. This, and the mobile version of the problem, also appear in R.P. Boas' super essay in [Hons3], Chapter 3. Euler's constant is discussed in [Simm]. My thanks to Dr. Scott Hunter for inspiration towards solving the convex polygonal wheel problem.

8 Möbius Madness: Tortuous Twists on a Classic Theme

8.1 Möbius Basics

One piece will result. As the top half strip of paper is attached to the bottom half strip, at A

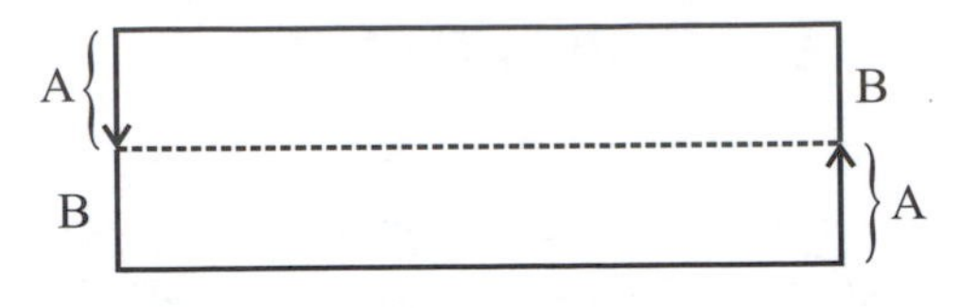

and again at B, the two half strips do not separate.

Taking it Further 1 Answer. Making two half twists (in fact, any even number of half twists) corresponds to gluing together the two ends of a strip of paper with orientation preserved. In this case, the end segments of the top half of the strip are glued together, and the same for the ends of the bottom half. Thus two separate (but intertwined) pieces result when the figure is cut along the center line. Notice that each piece is a Möbius band with the same number of half twists as the original band!

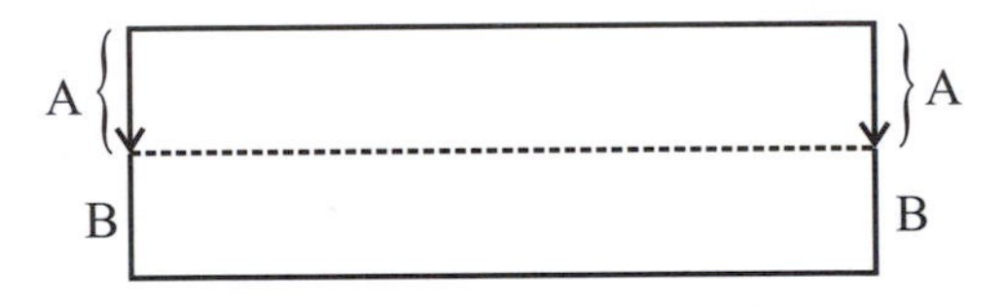

An odd number of half twists results in a reversal of the two ends of the original strip of paper. As before, one piece emerges when the band is cut in half.

Challenge. A Möbius band contains an odd number n of half twists. How many half twists will be present in the single loop of paper that emerges from cutting the band in half?

Taking It Further 2 Answer. Trisecting a Möbius band corresponds to the following diagram. When we cut along the dotted line, the top and bottom thirds connect to form one piece while the middle third is attached to itself and forms a second piece.

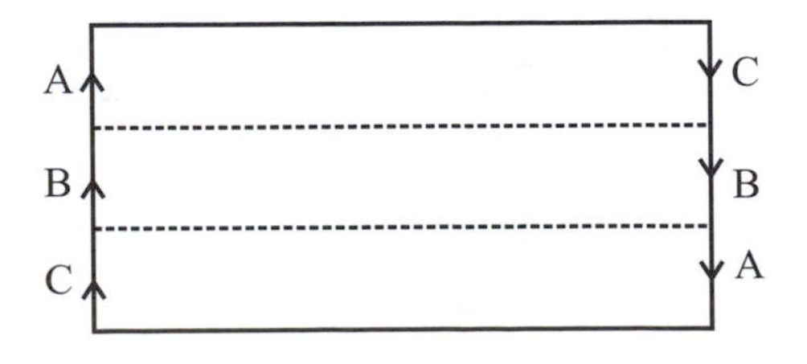

Cutting a band with five half twists into fifths results in three separate pieces.

Taking it Further 3 Answer. Tease the double-layered Möbius band apart to see that it is nothing more than a single band with four half twists. Thus, when it is cut in half, two pieces result—contrary to its single-layered cousin!

8.2 A Diabolical Möbius Construction

The connecting strip is attached at both ends to the same "half" (top or bottom) of the Möbius loop, unless one or both twist sequences contains an odd number of half twists. If odd, only one piece emerges when the figure is cut along the dotted line. Two distinct (intertwined) pieces result if each sequence of twists contains an even number of half twists.

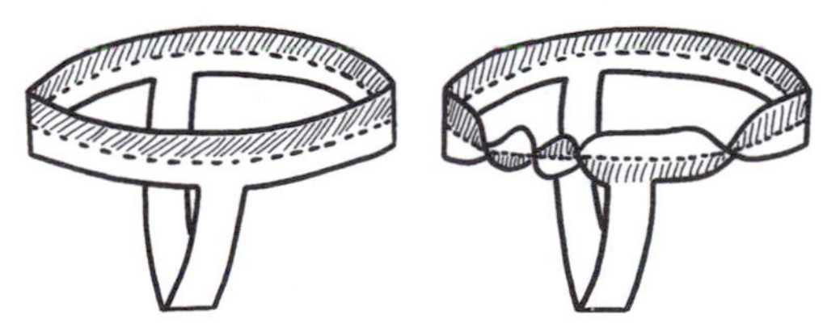

8.3 Another Diabolical Möbius Construction

Irrespective of the number of half twists, a flat square band of paper always results.

Challenge. What happens if *both* bands are given half twists?

Note. Can you see that the construction of this section is equivalent to the following variation of the one in section 8.2?

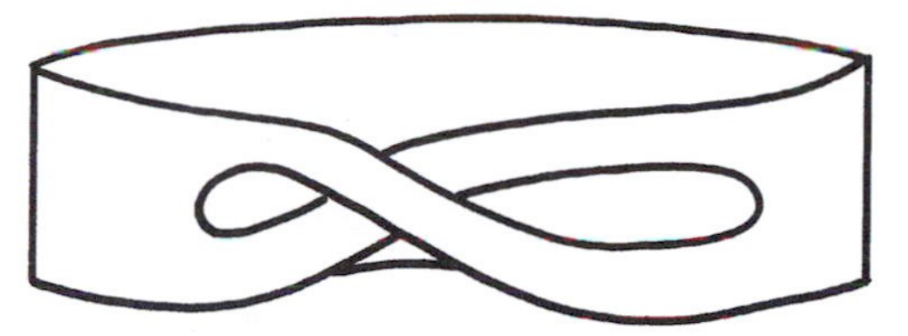

Acknowledgments and Further Reading

The Möbius band was discovered almost simultaneously, but independently, by the topologist/applied mathematician Johann Listing and the mathematician/astronomer Augustus Möbius in 1858. Although not mentioned in this section, the key feature of the Möbius band of interest to these gentlemen was its curious single-sided single-edged nature. The one-sidedness of the surface was later exploited by the B. F. Goodrich company in designing conveyor belts that last twice as long as conventional belts by uniformly spreading the "wear and tear" on both sides of the strip. A simple review of the Möbius band and some more illustrations of its practical use can be found in Theoni Pappas' super book [Papp]. Also see [Gard3], [Gard13] and [Hoff]. I first saw the diabolical Möbius construction of Section 8.2 in an exercise on surface classification in B. H. Arnold's wonderful book [Arno]. The second construction appears in [Gard22]. For some wildly tricky Möbius-like dissections have a look at [Gard6], Chapter 2.

9 The Infamous Bicycle Problem

9.1 Which Way Did the Bicycle Go?

A bicycle's back wheel always points toward the front wheel's point of contact with the

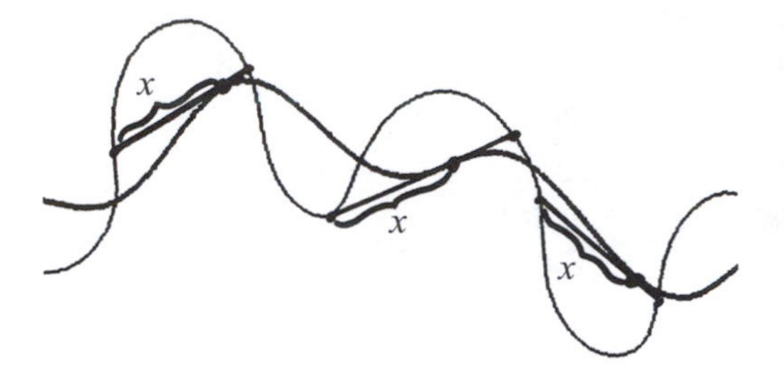

ground. So the direction of the back track at each point along the curve—more precisely, the direction of the tangent line at each point along the curve—must point toward the front curve. Also, since the bicycle is of constant length, the length of the tangent line segment that hits the front curve must be constant. This fails if we assume the bicycle travelled from left to right. It must be the case, then, that the bicycle travelled in the opposite direction. As a bonus, we also now know the length of the bicycle!

All the bicycles I have examined are three feet long from axle to axle. Could two bicycles of different lengths produce exactly the same tracks?

Hard Challenge. If the bicycle travelled in a straight line, this method would break down. It would be impossible to determine which way the bicycle went. This would also be the case if the bicycle travelled in the arc of a circle. The symmetry of that situation foils the above approach. Are these the only two cases for which the method fails? (Assume the bicycle never comes to a stop to turn a sharp corner.)

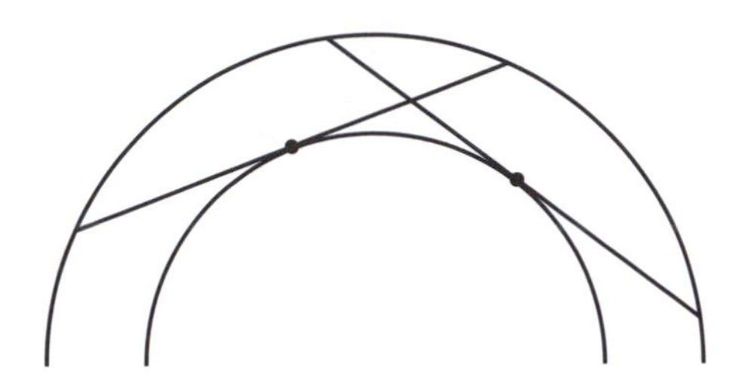

This is a very difficult question. If the bicycle travelled in a closed loop, and we could not determine which way it travelled, must that loop have been a perfect circle? If the bicycle came from and wandered off to infinity, and we couldn't deduce the direction of travel, must this path be a straight line?

9.2 Pedal Power

Surprisingly, the bicycle moves backward even though as you push the pedal it rises up toward you in a curious way that seems counter to the

A Note on Bicycle Tracks

Here is something curious. Let r be the distance between the points of contact of two bicycle wheels with the ground. First observe that if the back wheel moves in a perfect circle (with radius a) the front wheel does the same (a circle of radius b.) The region between the two tracks is an annulus whose area turns out to be πr^2—the area of a circle of radius r, independent of the radii of the two circles traced! That is,

$$\text{Area} = \pi b^2 - \pi a^2 = \pi(b^2 - a^2) = \pi r^2.$$

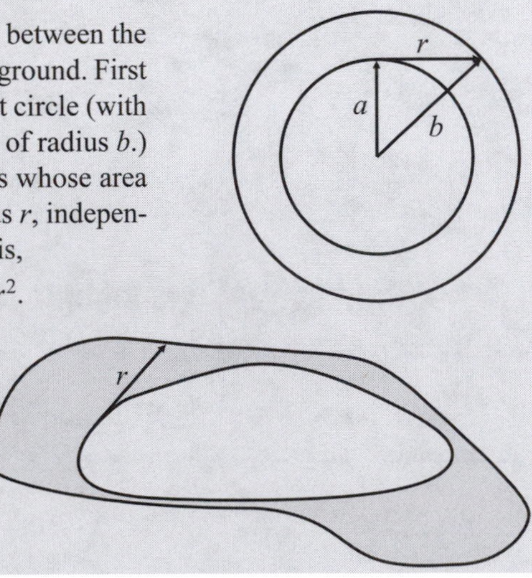

In fact, even if the back wheel traces a closed convex curve different from a circle, the area of the region between the two tracks traced is always πr^2, irrespective of the shape of the curve traced!

To see why this is true, first consider the case where the back wheel travels along the

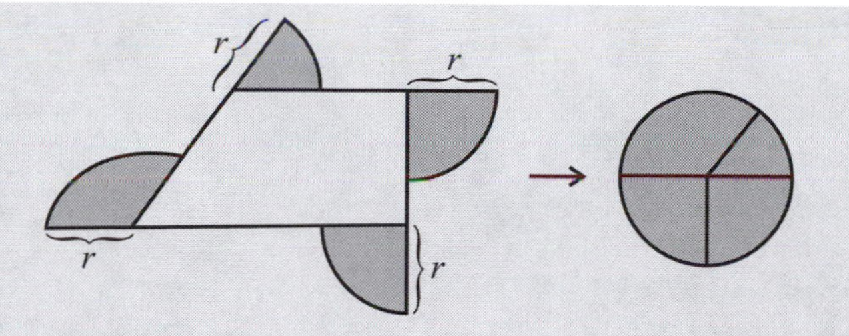

edges of a convex polygon, turning sharply at each corner (in fact, pivoting about the point of contact). The front wheel travels in straight lines as the back wheel follows the edges, and sweeps out sectors of a circle of radius r at each corner. These sectors fit together to form a complete circle of radius r, and hence the area between the two tracks in this polygonal case is πr^2.

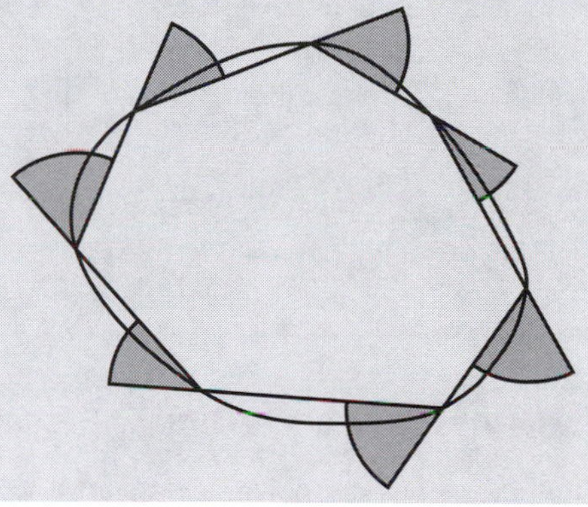

Any curve can be approximated by a polygonal curve and so, by a limiting argument, the area between the two tracks, even in a limit, remains πr^2.

This result was observed by Mamikon Mnatsakanian in 1959 and is explained more fully in [Mnat] (though not in terms of bicycles!) Also, in this article, he mentions the corresponding phenomenon for non-convex curves, curves with different winding numbers, and curves sitting in three-dimensional space.

nature of your push! (Try it!) A little physics explains what is going on.

In this schematic diagram, r is the length of the pedal shaft, L is the radius of the back wheel and F is the size of the force applied to the pedal in the backwards direction. Suppose the gear ratio is such that one revolution of the pedal causes the back wheel to rotate s times. Thus, if the force F initiates rotational motion with angular velocity ω on the pedal and its shaft, the back wheel rotates with angular velocity $s\omega$.

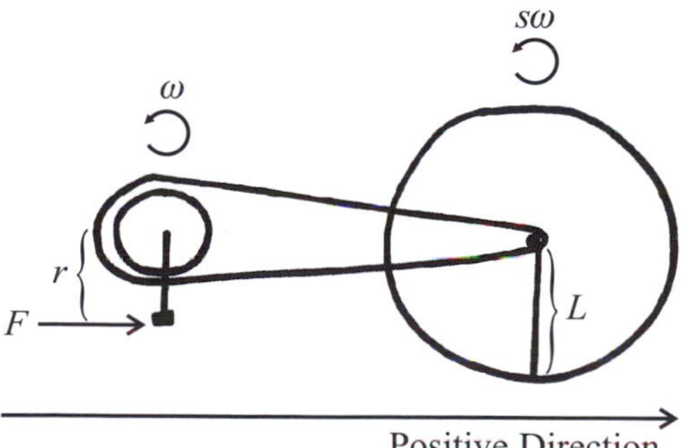

The tangential velocity of the point of the wheel in contact with the ground is thus $Ls\omega$. The pedal has velocity $r\omega - Ls\omega$ (not only does it have a tangential velocity, it moves with the bicycle!). A force applied to a point causes that point to move in the direction of the force, if at all, and so the pedal can only move in the backwards direction. Thus $r\omega - Ls\omega \geq 0$. That is, $(r - Ls)\omega \geq 0$. For normal bicycles, $r < L$ and $s > 1$ and so $(r - Ls) < 0$. It must be the case then that $\omega \geq 0$ and the bicycle moves backwards!

Question. What would it be like to ride a bicycle with its chain hooked on in a figure eight?

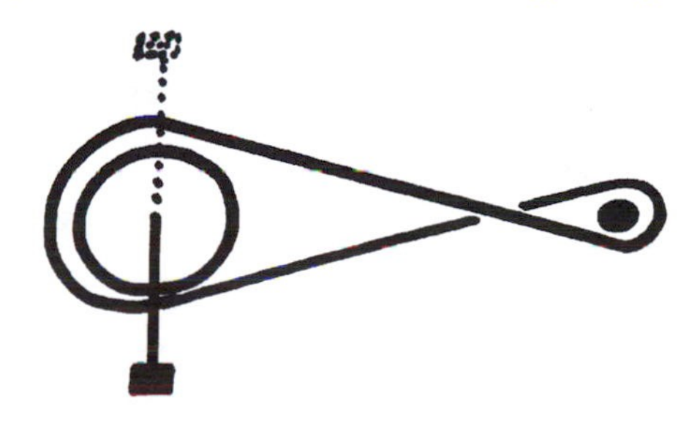

9.3 Yo-Yo Quirk

Surprisingly, the yo-yo moves in the direction of the pull and winds up the string in the process! One can apply analysis similar to the pedal problem of Section 9.2, or simply note that this system, at each instant, is somewhat analogous to that of pulling a balanced rod by a piece of string. Of course the rod is going to fall (move) toward you. The torque applied about the point of contact with the ground tends to rotate the stick or the yo-yo so as to make it roll towards the direction of the pull.

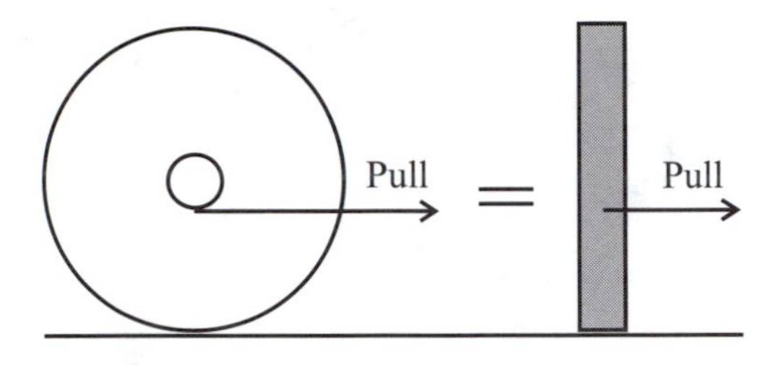

Acknowledgments and Further Reading

The bicycle track problem was first shown to me by John H. Conway in 1992 and has since appeared in [Konh]. I use this experiment in all my calculus classes when we first talk about tangent lines to curves. David Benbennick of the University of Alaska, Fairbanks has pointed out to me examples of *non-differentiable* curves for which it is impossible to determine in which direction a bicycle travelled. Have the back wheel of a bicycle follow the figure eight of two tangential circles of the same radius, for example. Here the bicycle must come to a balanced standstill at the point of intersection so that the rider can change the direction of the handlebars.

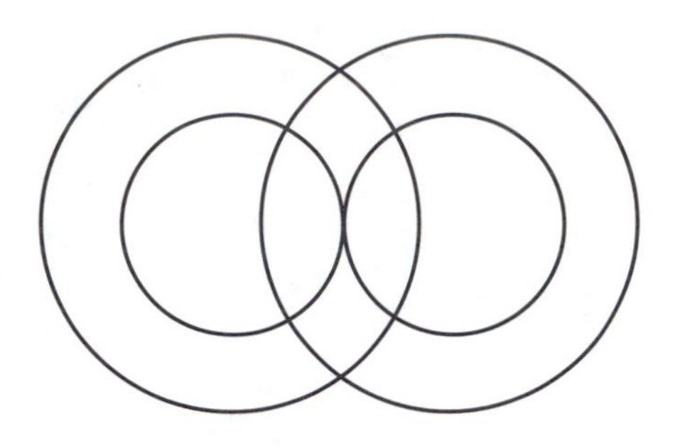

The bicycle pedal problem, also mentioned to me by John H. Conway, appears in [Gard17]. The solution to it presented here is due to the work of D. E. Daykin [Dayk].

10 Making Surfaces in 3- and 4-Dimensional Space

10.1 Making a Torus

Bend the cylinder with a half twist. This rotates pairs of edges B and C so that they can be glued together. The result is another torus! (See figure at the bottom of the page.)

10.2 A Torus with a Serious Twist

We are challenged to connect two ends of a cylinder from the same direction. This requires pushing one end of the tube through itself so as to link up with the other end. Doing this would create a peculiar surface, perhaps familiar to you already, the *Klein bottle.* Of course the con-

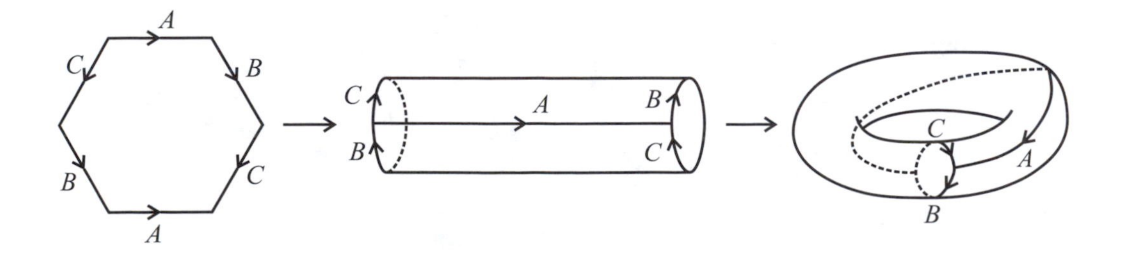

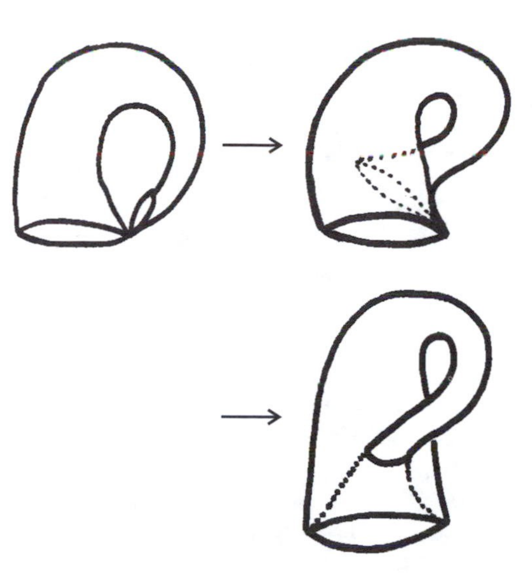

struction of such a surface is impossible in our three-dimensional world. (No wonder you were having trouble!) However, the Klein bottle is a perfectly valid mathematical object if situated in four-dimensional space! To see how, read the Note after section 10.3.

10.3 Capping Möbius

To sew a disc onto the boundary of a Möbius band, all we need do is choose a center point P anywhere in the fourth dimension and draw radial spokes from it to the boundary of the band. This will do the trick! (Read the next note to learn about the fourth dimension!)

A Note on the Fourth Dimension

I claim it is possible, in some abstract sense, to remove a pebble from inside a hollow rubber ball *without* cutting, tearing or puncturing the rubber. The way to come to grips with problems like these is to first consider the analogous problem one dimension lower. In this case, imagine a circle of string lying on a table top and a flat pebble captured inside. No matter how we push or stretch the circle on this table, the pebble always remains inside. To the inhabitants of a two-dimensional world, with absolutely no notion of looking up or down, the removal of the pebble is simply an impossible task. We, however, are three-dimensional beings and have the advantage of viewing this predicament from another dimension. To us, its solution is trivial. All we need do is lift up a small portion of the string in the direction of the third axis of motion, the z-direction, push the pebble along the table top through the small "opening" that has been created (to the inhabitants of the table top it looks like a hole in the wall has mysteriously appeared), and then bring the pinched portion back down from the third dimension to the flat table top to complete the task.

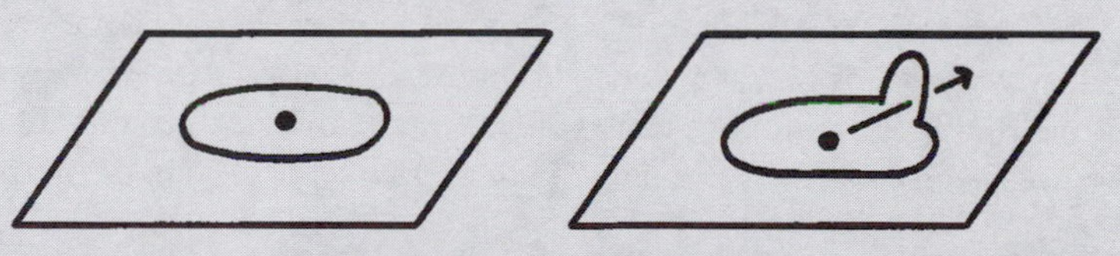

To remove a pebble from a hollow rubber ball, all we need do is stretch a small piece of the sphere a little way up into the fourth dimension to create what seems like a small opening, pull the pebble through this hole, and bring our small patch back down to the third dimension to reform our three-dimensional sphere. Simple!

Many people fret for hours (days, years!) as to what the fourth dimension could be. To a mathematician it doesn't really mean much. For example, a two-dimensional world such as the xy-plane, is simply described as the set of points

$$\mathbb{R}^2 = \{(x, y) : x, y \in \mathbb{R}\}.$$

The three-dimensional world is simply the set

$$\mathbb{R}^3 = \{(x, y, z) : x, y, z \in \mathbb{R}\}.$$

To obtain a four-dimensional world, just add another variable to the list:

$$\mathbb{R}^4 = \{(x, y, z, w) : x, y, z, w \in \mathbb{R}\}.$$

The mathematics of $\mathbb{R}^4$ works the same way as for $\mathbb{R}^3$ and $\mathbb{R}^2$. For example, we could regard a two-dimensional circle $C = \{(x, y) : x^2 + y^2 = 1\}$ as a three-dimensional object by writing instead $C = \{(x, y, 0) : x^2 + y^2 = 1\}$. Moving a point $(x, y, 0)$ on the circle one unit in the "z-direction" will give the point $(x, y, 1)$.

Similarly a sphere $S = \{(x, y, z) : x^2 + y^2 + z^2 = 1\}$ can be regarded as a four-dimensional object $S = \{(x, y, z, 0) : x^2 + y^2 + z^2 = 1\}$ and moving a point $(x, y, z, 0)$ on the sphere one unit in the "w-direction" gives the point $(x, y, z, 1)$. To a mathematician, removing a pebble from the interior of a sphere via the fourth dimension is no different from removing a flat pebble from the interior of a circle via the third. The fourth dimension gives us plenty of room to maneuver.

By the way, all sorts of neat things happen by moving to the fourth dimension. For starters, we can easily untangle all knots. Suppose we have a tangled and knotted loop of string. In three dimensions this would be impossible to unknot unless we pull out a pair of scissors and make a few snips here and there (actually, one would suffice). However, if we were to lift a small segment of this loop up into the fourth dimension, we would effectively create what seems like a small gap in the loop. We could then maneuver the string to untangle it and then bring our small segment back down from the fourth dimension to the third to give us our complete three-dimensional untangled loop—no snips involved!

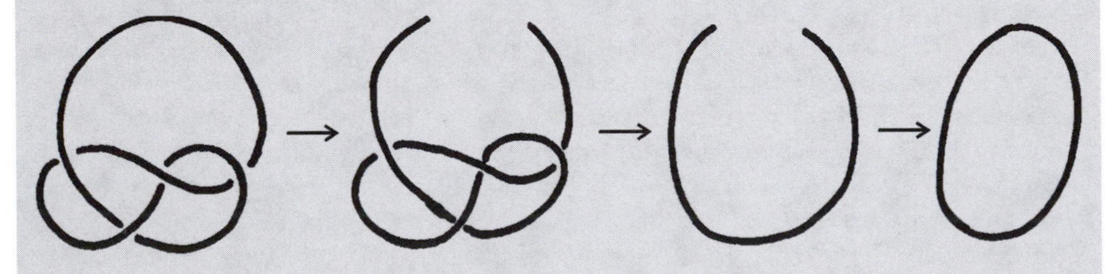

We can unlink two linked rings by the same trick (see section 17.2) Also, Klein bottles exist in four dimensions with no punctures and no self-intersections (see sections 10.2 and 17.2). The problem with the Klein bottle is that to form it we need to cut a hole in a tube so that we can connect the ends of that cylinder "from the same side." However, by lifting a small patch up to the fourth dimension we can avoid cutting altogether, effectively creating what seems like a hole to push our tube through. Many tricks are possible in the fourth dimension!

Challenge 1. A triangular card with vertices labelled 1, 2, and 3 in clockwise order lies flat on a table top. To obtain a card with vertices labelled in reverse order we simply flip it over. How

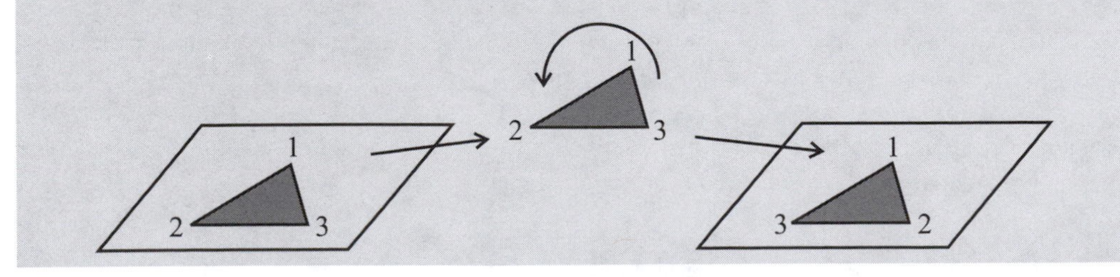

then could we turn a cardboard tetrahedron with vertices labelled 1, 2, 3, and 4 into its mirror image?

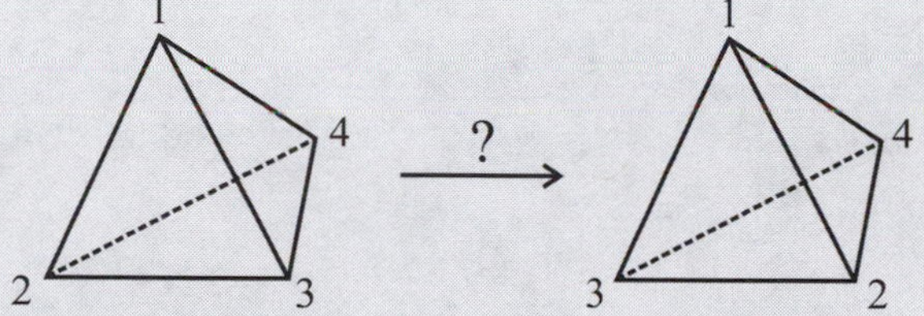

Challenge 2. Suppose a sphere in three dimensions falls through a two-dimensional world. The inhabitants of that world would first see a point that expands into a circle that grows, shrinks back to a point, and then disappears. What would we see if a sphere from the fourth dimension were to fall through our three-dimensional universe?

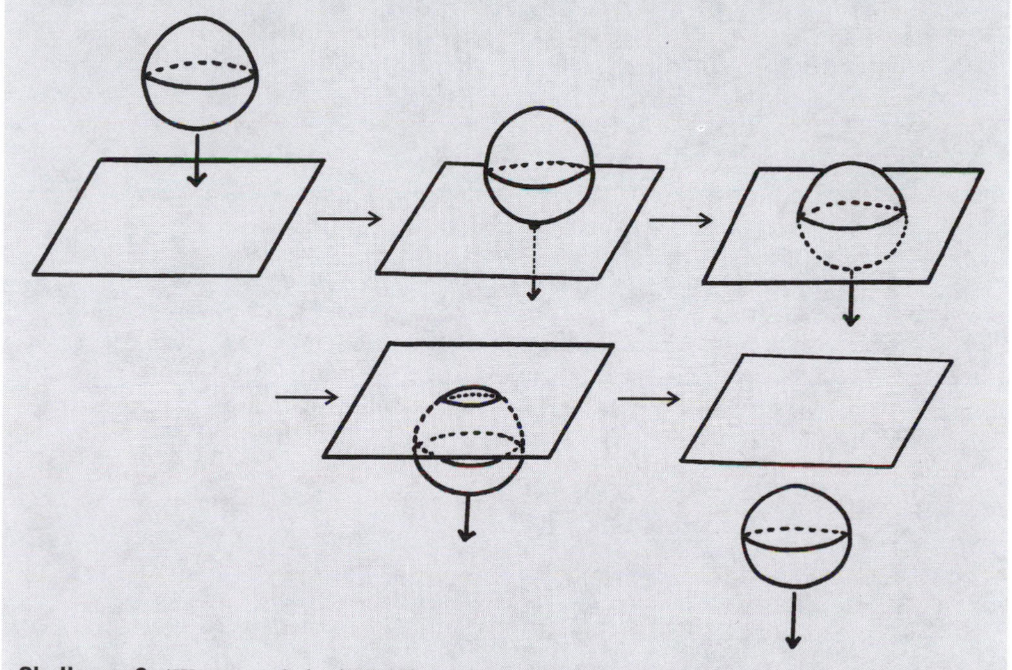

Challenge 3. (*Warning:* Calculus!) The volume of a sphere is obtained by taking the integral of cross-sectional areas. Each cross section in this case is a circle. Thus,

$$\text{Volume} = \int_{-r}^{r} A(x)\,dx = \int_{-r}^{r} \pi\left(\sqrt{r^2 - x^2}\right)^2 dx$$
$$= \frac{4}{3}\pi r^3.$$

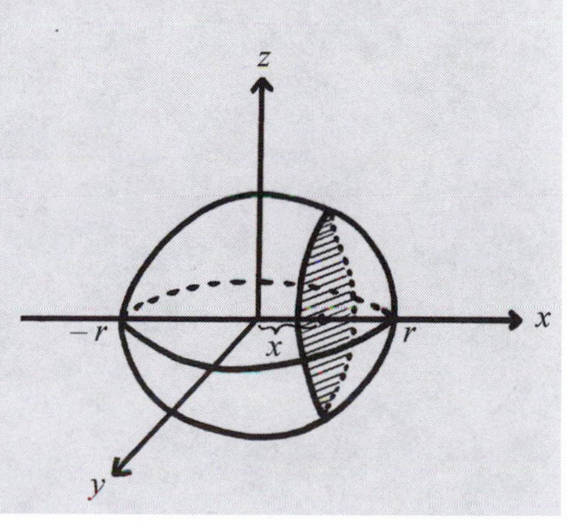

The volume of a four-dimensional sphere is computed in the same way: It is the integral of the cross-sectional volumes, all of which are spheres. Show that the volume of a four-dimensional sphere of radius r is

$$V = \frac{1}{2}\pi^2 r^4.$$

What is the volume of a five-dimensional sphere?

A Note on Constructing Surfaces

One can form surfaces from even-sided polygonal pieces of paper identifying pairs of edges. So far we have seen the construction of a torus from a square and from a hexagon, and a Klein bottle from a square.

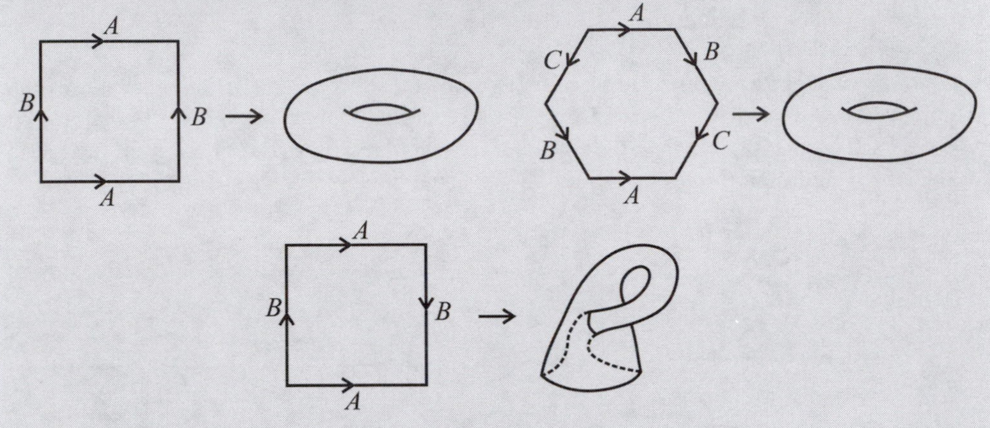

The surface in section 10.3 also arises in this way. Consider the square with edge identifications illustrated to the right. I claim this is nothing more than a Möbius band with a disc sewn onto its boundary! To see why, cut out a strip from the center of the square. When glued appropriately, this strip forms the Möbius band; the other two pieces, when glued, form the disc. To undo the damage of our cutting we now must reattach the disc back on to the Möbius band. This is the challenge of 10.3.

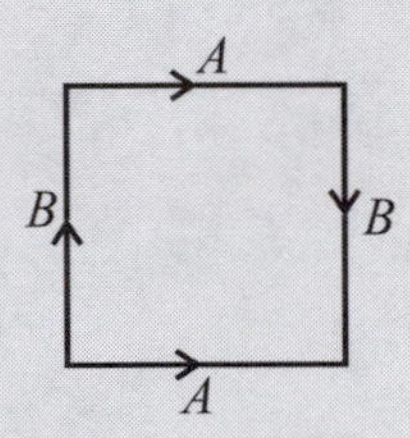

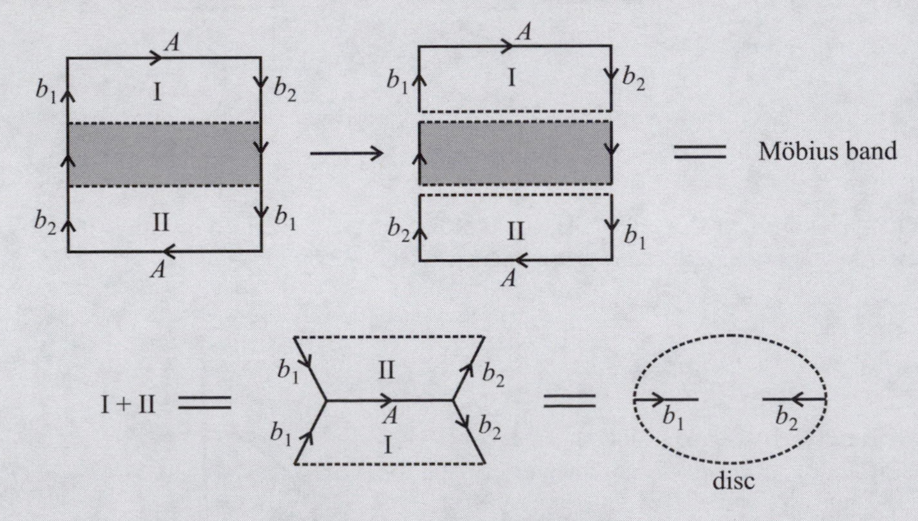

Challenge 1. Do the same trick to show that a Klein bottle is just two Möbius bands sewn together!

Challenge 2. Other pairs of edge identifications on a square are possible. What surfaces do they yield?

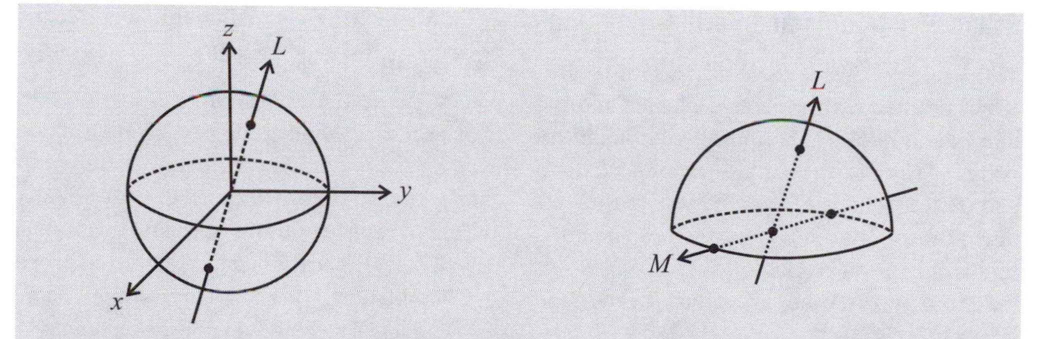

The surface arising in section 10.3 is called a *real projective plane*. There is yet another way to envision it. Let S be the set of all lines in three-dimensional space that pass through the origin. To identify such a line we need only indicate at which two points it intercepts a sphere (see the figure on the left above). Actually we need work only with one hemisphere and indicate the one point at which a line intercepts the hemisphere (unless the line happens to intercept the equator, in which case two opposite points on the equator represent the same line). (See the figure on the right.)

Thus S is equivalent to a disc with pairs of opposing boundary points identified as one point, which in turn is equivalent to a square with opposite points identified. The set of all lines passing through the origin in three-dimensional space is thus precisely the real projective plane!

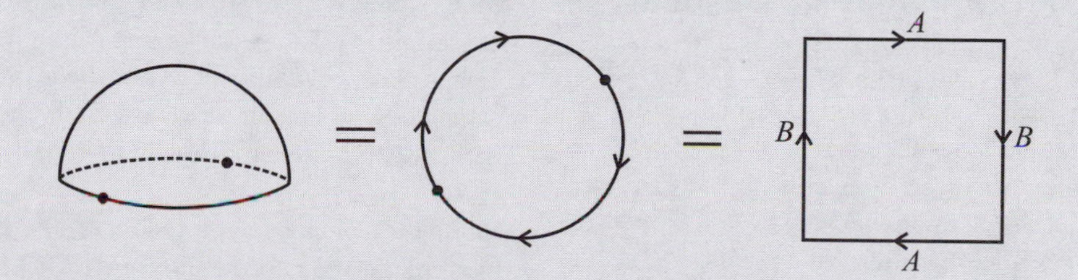

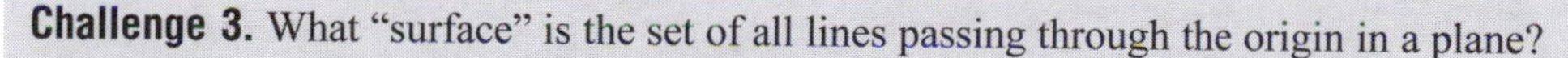

Challenge 3. What "surface" is the set of all lines passing through the origin in a plane?

Challenge 4. Show that a double-donut results from identifying the opposite edges of an octagon.

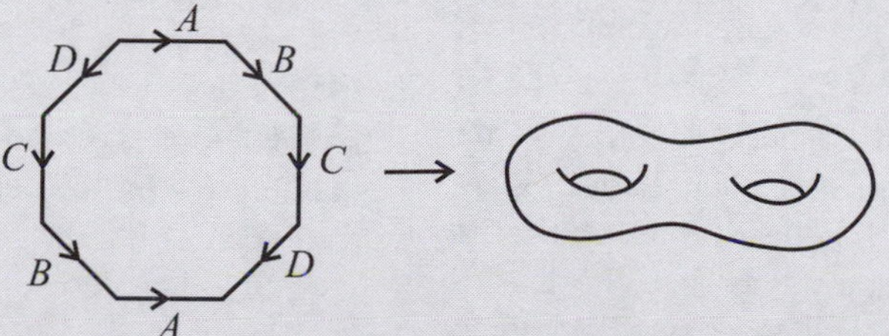

Mathematicians have proven that identifying edges of even-sided polygons always creates surfaces and, moreover, have classified all the surfaces that arise in this way. It turns out that the sphere, the torus, and the real projective plane are the building blocks of these *compact* surfaces (loosely speaking, those surfaces composed of a finite amount of material and having no open edges). The double-donut is built as the join of two single tori and the Klein bottle is built, surprisingly, as the join of two real projective planes! To learn more about this classification theorem see [Firb], [Fran2] or [Mass]. (*Warning:* These are advanced books.)

Acknowledgments and Further Reading

To learn more about the construction of tori, Klein bottles, real projective planes, and the like have a look at the thoroughly delightful books [Arno], [Firb], [Hild], [Jaco], chapter 4, and [Pete3], chapter 2. Also check out my article [Tant1]. Martin Gardner writes about the fourth dimension in [Gard6], chapter 4. See also the classic text [Abbo] as well as [Gard7] and [Ruck].

11 Paradoxes in Probability Theory

11.1 The Money or the Goat?

You may argue that after the game show host acts, two doors remain, one containing the prize, the other the goat; so the chance of selecting the correct door, either by sticking with the chosen door or switching, is 50% either way. Thus switching offers no advantage. But surprisingly, this reasoning is incorrect! There is an advantage to switching doors.

The initial selection of a door fits one of two scenarios: Either the contestant has selected the correct door or she has not. Let's examine each scenario in turn.

Scenario 1: (The first guess is correct.) The host's decision of which door to open is arbitrary and has no effect on her choice. If the contestant knows she has chosen the correct door, she will stick with her choice in order to win the prize.

Scenario 2: (Her first guess is incorrect.) In this case the host's action is not arbitrary. He has no choice as to which door to open (did you notice this when you simulated the experiment?) and in effect has revealed precisely where the prize lies (behind the door he didn't open). Thus, within this scenario the contestant should switch.

Of course the contestant does not know which scenario applies and hence which action to take. But which is more likely? There is a $1/3$ chance that she chose the correct door initially and a $2/3$ chance she did not. Thus she must reason that scenario 2 is more likely and hence she should switch her choice of doors! There is a $2/3$ chance she will win doing this (and a $1/3$ chance she won't). You can check these figures for yourself by conducting the experiment again several times, always choosing to switch in the process.

Here's a second and more succinct analysis of the game. Suppose the contestant adopts a "never switch" strategy. Then she will win the prize with probability $1/3$ since there is a one-third chance she will initially select the correct door. On the other hand, with an "always switch" strategy, the contestant will win whenever she initially chooses an incorrect door. This happens with probability $2/3$. Thus she is twice as likely to win with an "always switch" strategy than with a "never switch" one.

Challenge 1. Suppose you face n doors, k of which contain the same fabulous prize. In playing the analogous game, the host reveals r unsuccessful doors, $1 \le r \le n - k - 1$, after you have made your initial choice. Is it always to your advantage to switch choice of doors?

Challenge 2. Suppose, in the three-doors game, Monty Hall forgets behind which door the prize lies and purely by luck opens a door that reveals a goat. Is it still to the contestant's advantage to switch doors?

Challenge 3. Here's another puzzler to test your probabilistic reasoning skills. One bag contains a red ball and a white ball, and another holds two white balls. You are handed one bag at random. Reaching in, you pull out a white ball. What is the probability that the other ball in that bag is red?

11.2 Double or ... Double!

This paradox still disturbs philosophers and mathematicians. It is related to the Wallet Game described by Martin Gardner in [Gard12, p. 106], in which two people each count the amount of money in their wallets. Whoever has the smallest amount of cash will win the contents of the other person's wallet. Both people could correctly reason that they have more to gain than they could lose, so the game is definitely in their favor. But a game cannot simultaneously be advantageous to both players, and therein lies a paradox.

Martin Gardner hints that the resolution of this, and the Tootsie Roll® paradox as well, lies in questioning the assumption that there is a 50–50 chance that the other wallet or bag is "better" than the first. I agree. For example, in the Tootsie Roll® game I would certainly want to play if the bag I open contains only one Tootsie Roll®. I would reason that it is very unlikely my colleague would place just half a piece of candy in the other bag, and so I would almost certainly gain by switching bags. At the other end of the spectrum, I would know that I should "stay" with a bag groaning with 684 Tootsie Rolls®, for the other bag certainly cannot hold 1368 of them!

Similarly, in the Wallet Game I would want to know the statistics on the average amount of cash carried by folk of certain professions, on certain days of the week, and the like. Is it always a 50–50 chance that another person would have more money in their wallet than me? I doubt this would be the case if on one (very lucky) day I happened to be carrying $2041 with me! Laurence McGilvery resolves the Wallet Game paradox [McGi] by questioning the assumption that one would even *want* to play the game. I would play if I had just one cent in my wallet, but would not if I had $2041!

Challenge. Let's take the Tootsie Roll® paradox a little farther. Imagine you are trying to set up the game for a friend to play. Knowing that your friend is aware of the carrying capacity of the bag, and that you will never use fractional pieces of candy, is it possible to set up a two-bag game for which your friend could never "know" which choice to make, especially if she knows this is your mission?

For example, you would never place just one candy in one bag and two in the other (let's call this a "1–2 game"), for if your friend, by chance, chose the bag containing one candy, she would immediately know to switch. By the same reasoning, you would never want to set up an n–$2n$ game with n odd. You would never want to set up a $2n$–$4n$ game either, n odd, for if your friend by chance first chose the $2n$ bag, she would deduce that she should switch, since she knows you will never set up an n–$2n$ game. By the same reasoning, you will never set up a $4n$–$8n$ game, n odd, either; nor an $8n$–$16n$ game, a $16n$–$32n$ game, and so on. This rules out *all* possible choices of numbers! Your friend, given your strategy, will always know what to do: Switch!

Realizing this, you decide to put 16 pieces of candy in one bag and 32 in the other. Have you confused your friend after all?

11.3 Discord among the Chords

The problem here lies in defining what we mean by "select a chord at random." The three different calculations rest on three different interpretations. The first assumes that a chord is chosen by selecting a point on the circumference of the circle (the other being fixed); the second assumes that a chord is selected by choosing a midpoint; and the third chooses a chord by selecting a midpoint on a fixed vertical line. All approaches are valid—but they work with different *measures* on the space of all chords of a circle and thereby yield different answers. (See the Note on the next page.)

11.4 Alternative Dice

Label the first die 1, 3, 4, 5, 6, 8 and the second 1, 2, 2, 3, 3, 4. One can easily check that these dice behave, as a pair, the same way as ordinary dice. (What would it be like to play *Monopoly* with them?)

A Note on Probability and Measure Theory

Consider a simple probability calculation: *Determine the chances of obtaining an even number when rolling a die.* To compute this quantity we simply count the number of desired outcomes (the rolling of a 2, 4, or a 6; three in all) and divide by the total number of possible outcomes (six). We then (naively) define the *probability P* of rolling an even number to be the ratio of these two counts:

$$P = \frac{\text{Number of acceptable outcomes}}{\text{Total number of possible outcomes}} = \frac{3}{6} = \frac{1}{2}.$$

Note: We have assumed each outcome to be equally probable. Thus we have relied on the notion of probability to define what we mean by probability! This is a serious flaw with this naive method. This approach also assumes we are able to count, or *measure* in some way, the sizes of different sets. In some cases this is easy, as with the roll of a die, but others can be more complicated. For example: *what is the probability that a point chosen at random in the unit square* $[0,1] \times [0,1]$ *lies under the parabola* $y = 4x(1-x)$? To answer this we need to "count the number of points under the parabola and compare this to the total number of points in the square." This is, of course, meaningless!

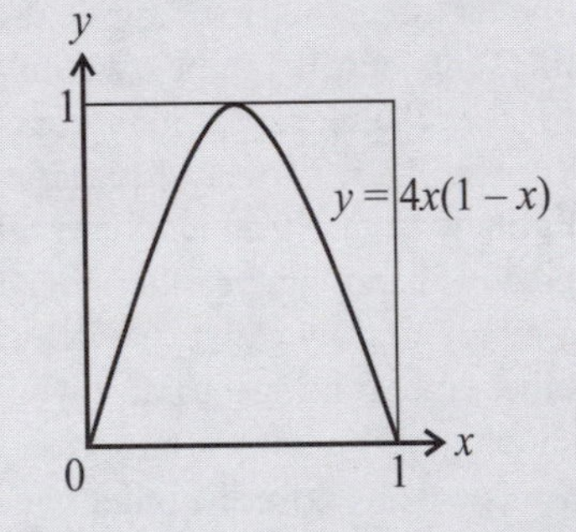

We are left with the challenge of extending our naive counting definition to a "non-countable" situation. Of course we just want to measure the area under the curve as illustrated. But for probability theory we want to be precise and careful about what we mean by "area." Are we really capturing the notion of "counting?"

In some sense the notion of area is just an artifice of human thinking. Mankind has declared the area of a square to be its side length squared(!) and the area of a rectangle its length multiplied by its width. These seem reasonable things to say, but there is no absolute reason as to why the areas of these shapes must be defined this way. Nonetheless, we need a starting point and on an intuitive level these definitions seem a good place to begin. From them we deduce that the area of any triangle is half its base times its height; the area of a polygon is the sum of the areas of triangles that subdivide it, and so on. By a limiting process, we can calculate the area of curved figures as well. (Look at the principle of Riemann integration: Subdivide into rectangles, sum the areas, and take a limit!)

Measure Theory attempts to distill from this process the key ideas that lead to a successful description of area. It assumes that we know how to measure the area of some basic or fundamental regions, sets, or quantities (in the case above, rectangles in a plane). From these basic sets we can build up new sets (via unions, intersections, and a simple limiting process) whose areas we can measure. If A is a measurable set we let $m(A)$ denote the measure of that set. The basic axioms of measure theory follow:

1. The empty set is measurable. It has measure zero.
2. If A and B are measurable sets and $A \subset B$, then $B - A$ is also measurable.
3. If $A_1, A_2, A_3, \ldots$ are a pairwise disjoint collection of measurable sets (that is, $A_i \cap A_j = \varnothing$ if $i \neq j$), then their union is measurable and

$$m\left(\cup_{i=1}^{\infty} A_i\right) = \sum_{i=1}^{\infty} m(A_i).$$

The development of measure theory allowed Lebesgue to completely revolutionize and generalize the notion of an integral. It also gave the means to put probability theory on a sound footing and resolve many of its paradoxes—including Bertrand's paradox.

Consider the space of chords in a circle. How do we measure the size of the set of chords that are longer than the side length of an inscribed triangle? We first need to explain which measure we wish to use. What are the fundamental sets of chords on which to build up a measure theory?

It turns out that Jennifer, Bill, and Joi of section 11.3 each based their measure theory on different fundamental packets of chords whose measures are assumed known. Jennifer worked with a small basic packet consisting of chords bunched together at one end point. She declared the measure of such a fundamental packet to be the fraction of the perimeter swept out by these chords at the other end. Bill's fundamental packet consisted of chords whose midpoints cluster within small regions of the circle. He declared their measure to be the fraction of the area of the circle taken up by these small regions. Joi had packets of horizontal chords whose areas are given by the fraction of the vertical line they cover. It's no wonder then that they each developed distinct final answers in analyzing this problem! The original problem was ill defined, for it never stated which measure to use. Take care with this when working in probability theory!

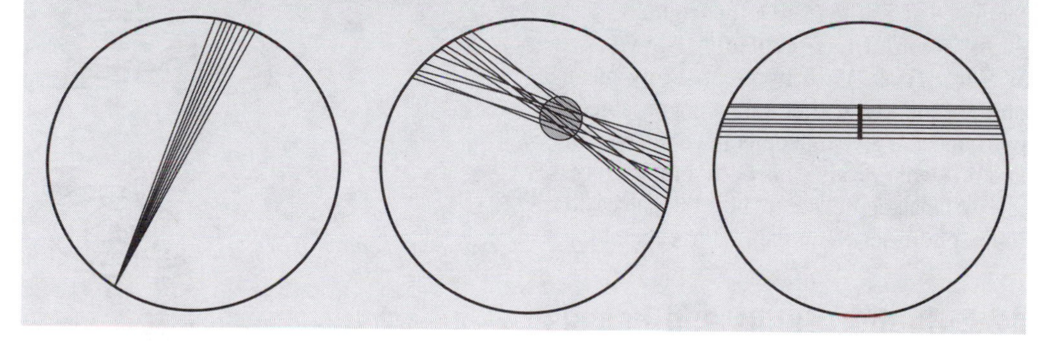

Taking it Further Answer. Tetrahedral dice can be relabelled 1, 2, 2, 3 and 1, 3, 3, 5.

Challenge. What about octahedral, dodecahedral, and icosahedral dice?

Acknowledgments and Further Reading

The Monty Hall problem is still worthy of discussion and experimentation; check out Luis Fernandez and Robert Piron's article [Fern], for example. Discussion of Bertrand's paradox can be found in Mark Kac and Stanislaw Ulam's simply wonderful book [Kac] and Charles M. Grinstead and J. Laurie Snell's fabulous probability theory book [Grin] (Monty Hall is here too!). E. T. Jaynes presents a detailed analysis of the paradox [Jayn]. See also [Gard3], [Gard4], chapter 19, [Gard13], and [Gard16] for other probability paradoxes. George Sichermann was the first to discover the alternative labelling of cubic dice. To learn more about these and other platonic solid dice have a look at [Brol] and [Gard18], Chapter 19.

12 Don't Turn Around Just Once

12.1 Teacup Twists

First imagine a strip of paper attached to the teacup. Given two full rotations, it is easy to see that the strip can be maneuvered around the cup back to its original status. Manipulat-

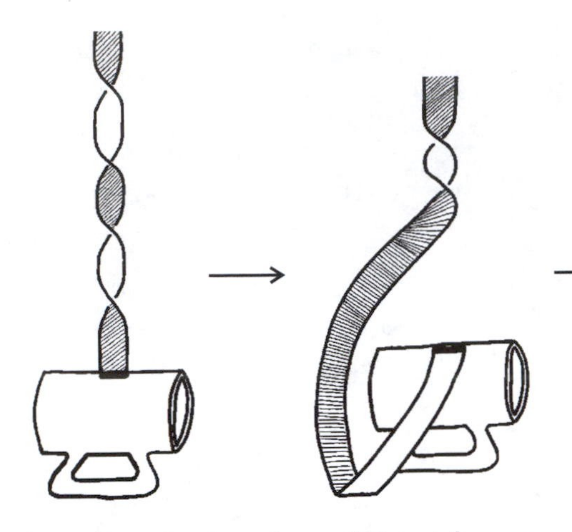

ing a line of strings is no different from manipulating a strip of paper. Either move the strings as a group as indicated above, or move them one at a time from left to right.

A variant of this problem is known as the Waiter's Trick. With two feet firmly planted on the ground a waiter balancing an object in one hand can give that object two full rotations and *not* twist or knot his body in the process. Alex Alapatt demonstrates how this can be done in the photo series on the facing page.

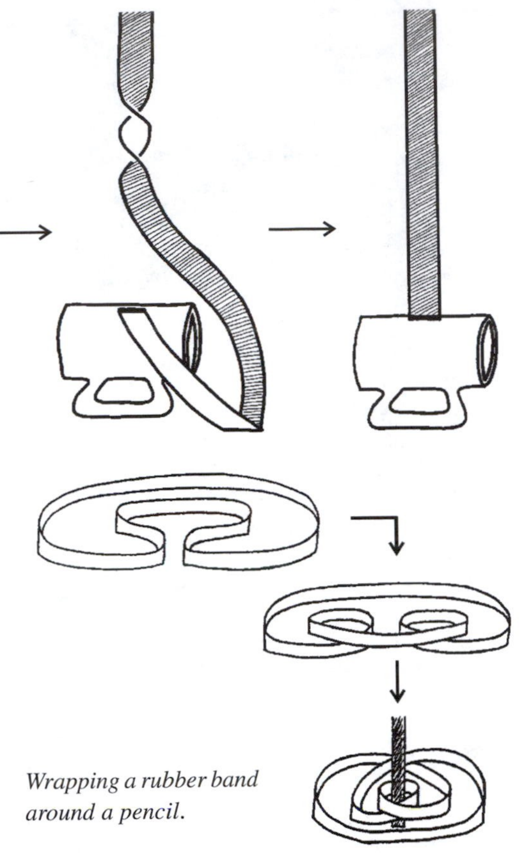

Wrapping a rubber band around a pencil.

12.2 Rubber Bands and Pencils

There is only one way to produce loops in a band of paper that is sitting on a table top while maintaining its vertical "walls." Pinch a fold of paper into the center of the loop. Pick up the end of this fold and flip it back over to the outer edge of the band to form two loops (with upright walls) within the band.

Any loop you produce in the band must be balanced by another loop counteracting the effects of the production of the first. For this reason, two loops appear in the procedure illustrated; and, in arbitrary manipulations, only even numbers of loops can appear. Adjusting the paper (or the rubber band) so as to "stack" these loops along with the original circuit of paper thus produces an *odd* number of loops to wrap around a pencil. And conversely, given any specific odd number, one can clearly wrap a band around a pencil that many times.

Challenge. The procedure above reminds me of a teaser you can try with friends. You will need a long piece of string and a teacup. (Use the one from problem 12.1!) Push the string through the handle of the teacup and tie a knot as shown. Tie the two loose ends of the string to a piece of furniture. Is it possible to remove the teacup from its knotty predicament *without* cutting the string?

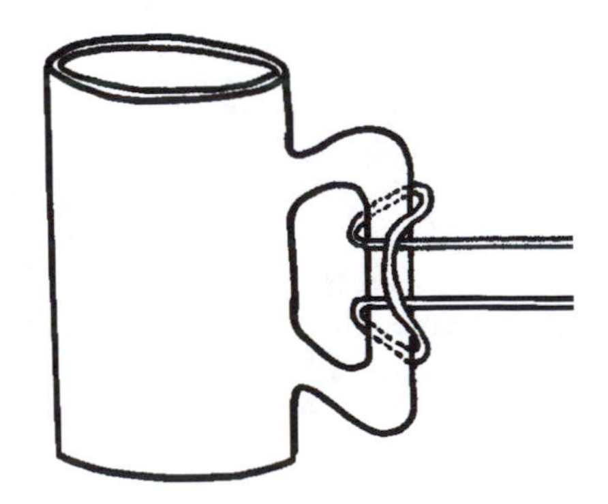

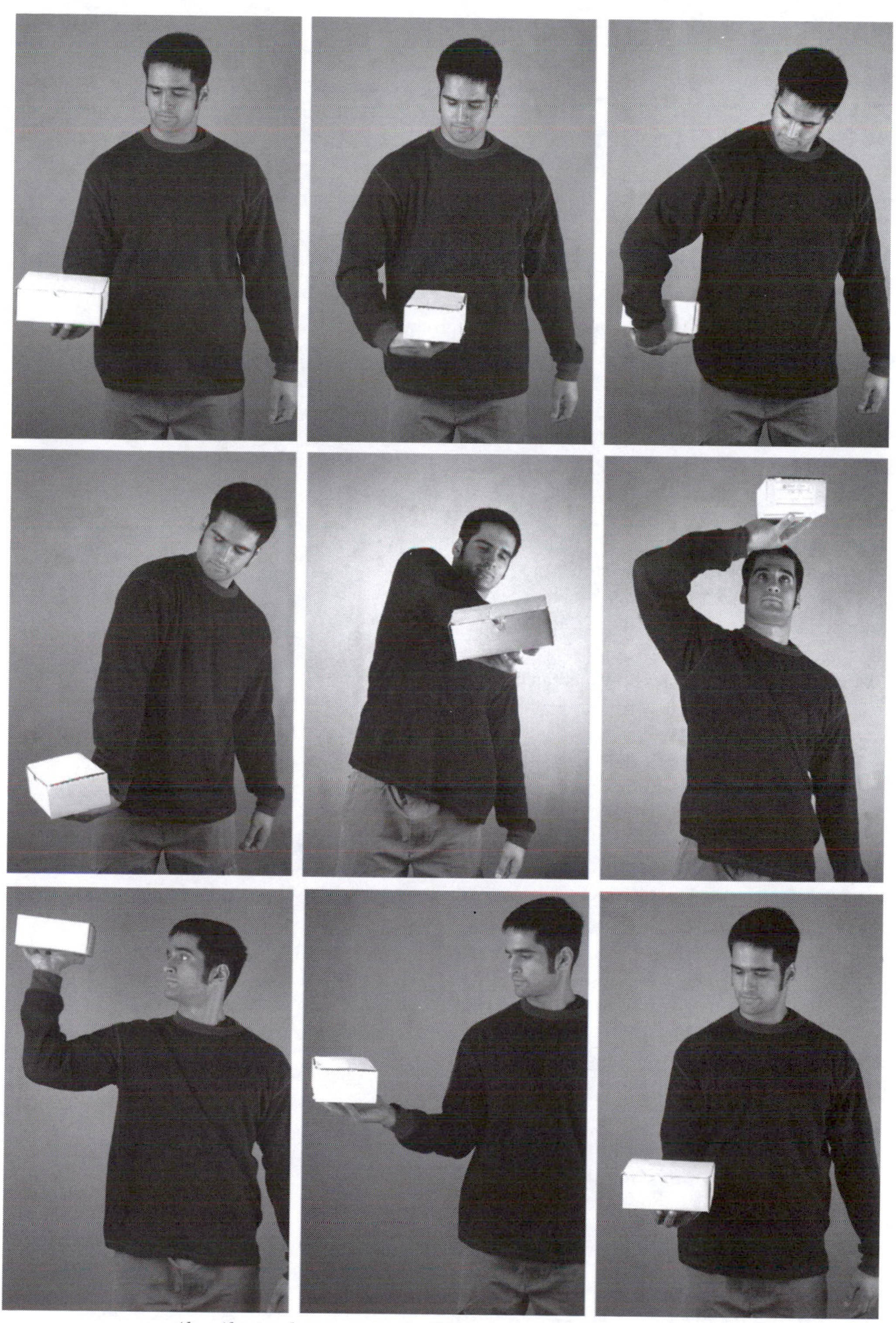

Alex Alapatt demonstrates the Waiter's Trick with a cardboard box.

Acknowledgments and Further Reading

My thanks to Stephen Gildea for alerting me to the rubber band problem. "Rubber band mathematics" and quantum physics have a deep connection. Look at D. Finkelstein and J. Rubinstein's paper [Fink]. Paul Dirac used the Waiter's Trick to demonstrate why angular momentum can come in packets of half a quantum, but of no other fractions—it is because two rotations of a body about an axis are equivalent, in a certain sense, to no motion at all. These issues are discussed in [Gard2], chapter 2 and [Newm]. See also chapter 5 of this book. For those advanced in their mathematical knowledge Glen Bredon's book [Bred] offers a complete mathematical analysis of the group of rotations in 3-space.

13 It's All in a Square

13.1 Square Maneuvers

You probably noticed that each circuit contains an even number of people. This is necessarily the case. In order for someone to return to the original empty square, the number of people taking a horizontal step to the right must be balanced by the same number of people taking a horizontal step to the left. Similarly for those taking vertical steps upward and downward. Thus the number of people involved in any circuit must be even.

The rules of this game mean that only an even number of people will ever be able to move. As a 5×5 grid contains an odd number of squares, this puzzle is impossible to solve! You must allow at least one person to move two squares or take a diagonal step to resolve the dilemma. (In fact, just one such person will do. Can you see why?)

The Note on the next page has an even swifter solution to the problem.

Challenge. If *everyone* is required to take two steps in a vertical or horizontal straight line, then the above reasoning again shows that the puzzle is unsolvable. But what if everyone is required to take a diagonal step? Is the puzzle solvable in this case?

13.2 Path Walking

Color the grid in the standard checkerboard scheme. All paths move through the black and white cells in an alternating fashion. As there are 25 black cells and 24 white, any path must begin and end on two distinct black cells. Thus no loop of steps nor any path commencing on a white cell can possibly visit each and every cell precisely once!

It is straightforward to check that desired paths emanating from any chosen black cell actually do exist.

Challenge 1. Develop a general scheme to show how to construct a path with the desired property emanating from any "black" cell in an arbitrary $n \times n$ array (n odd). Can you develop a scheme for cubical $n \times n \times n$ blocks (n odd)?

Challenge 2. Analyze path-walking possibilities on hexagonal triangular arrays.

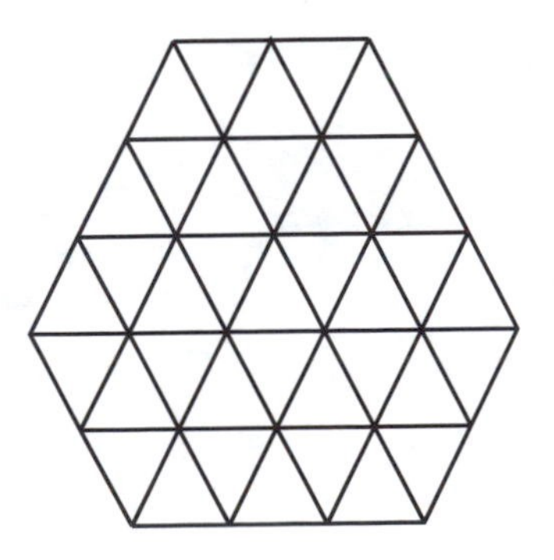

A Note on Modified Square Maneuvers

The discussion of section 13.1 showed that it is impossible for each person in a 5 × 5 grid to simultaneously take a single step in either a vertical or a horizontal direction to form a new configuration of 25 people standing in a square grid, one person per cell. However, if the person in the top left corner stands still it is possible for the remaining 24 people to complete such a maneuver.

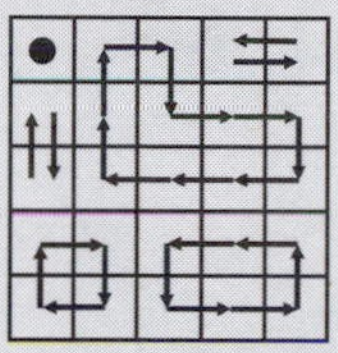

Does everyone have this option of standing still and allowing the remaining 24 people to shift places around them? For example may the person standing one cell over from the top left square remain still?

Yes	?	Yes	?	Yes
?	Yes	?	Yes	?
Yes	?	Yes	?	Yes
?	Yes	?	Yes	?
Yes	?	Yes	?	Yes

Try experimenting with having different folks stand still. You will soon suspect that not everyone has this option. If you experiment diligently, you may notice that the locations of people who admit solutions around them seem to form a checkerboard pattern across the grid. This is in fact the case!

Imagine the grid colored black and white in a checkerboard scheme. Those standing in a black cell will move to a white cell and vice versa. Thus, among the people moving, there must be an equal number of black and white cells. As this diagram has 13 black cells and 12 white cells, only those people standing in black cells have the option of remaining still. (This observation leads to a swifter solution to the original problem in section 13-1, for clearly not everyone can simultaneously move!)

Now draw a path from the top left cell to the bottom right cell as shown.

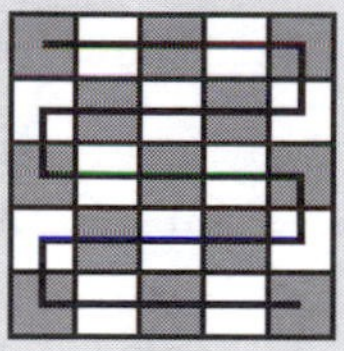

Any person standing in a black cell breaks this path into two pieces of even length (one piece may be empty). Imagine pairs of people swapping places along the length of each trail segment. This shows that it is indeed possible for a person in a black cell to stand still while the remaining 24 folks interchange places.

Challenge 1. A crowd of 125 people stands in a 5 × 5 × 5 cubical lattice. Each person is allowed to move one cell over: above or below, to the left or right, or to the front or back. Who may remain still and let the other 124 people interchange places?

Challenge 2. Now 25 people stand in a 5 × 5 square grid. Each person is allowed to move to a new cell in the way a knight moves across a chessboard—two squares in one (vertical or horizontal) direction and then one square over in an orthogonal direction. It is impossible for everyone to complete this maneuver simultaneously. (Why?) However, if one person remains still, is it possible for the other 24 people to complete the maneuver around her? Which people have the option of remaining still?

13.3 Square Folding

If, as a result of your devilish folding scheme, one square is facing upward, then its neighboring squares face downward. Thus the squares that face upward and downward form a checkerboard pattern across the original square. This pattern could be reversed for a friend doing the experiment along with you. I simply ensured that the two sets of numbers that alternate across the grid both sum to 63!

U	D	U	D
D	U	D	U
U	D	U	D
D	U	D	U

Challenge. Devise a similar folding trick for triangular sheets of paper.

Acknowledgments and Further Reading

I have written about these grid-walking puzzles [Tant3], [Tant4]. The Square Maneuver problem is a classic one and also appears in [Vaki]. The Square Folding problem appears in [Gard21].

14 Bagel Math

14.1 Slicing a Bagel

Slice the bagel along the diagonal as shown. Believe it or not, this produces two perfect intersecting circles! (If you don't believe me,

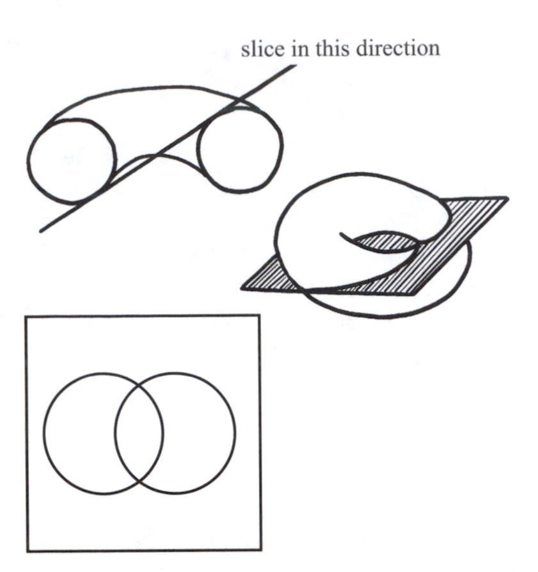

check out Wolf von Rönik's article [Röni] for the mathematics of this.)

14.2. Disproving the Obvious

Curves that completely circumnavigate the bagel in either direction, or both, fail to divide the surface into two distinct regions. These curves have no inside or outside as it were.

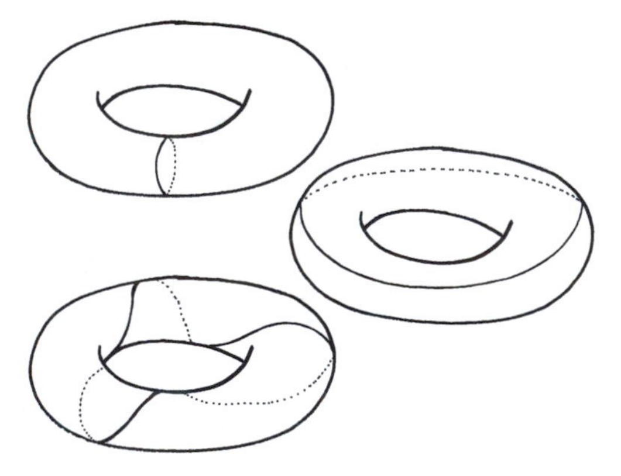

14.3. Housing on a Bagel

Indeed there is a toroidal solution!

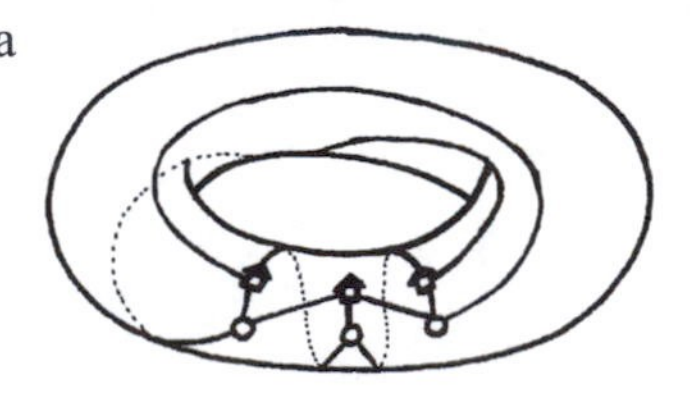

Challenge. Does a solution exist on a Möbius band?

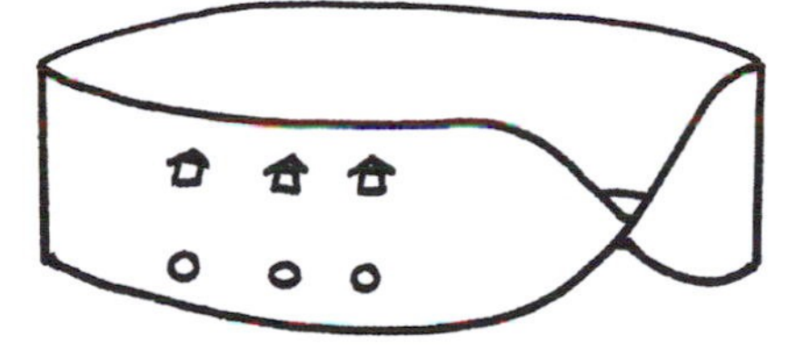

14.4. Tricky Triangulations

Let t be the total number of triangles used in a given triangulation, e the number of edges, and v the number of vertices. Since every triangle has three edges and each edge is counted twice, $3t = 2e$. Therefore t cannot be odd.

To show that any triangulation of a torus must involve at least 14 triangles, we use Euler's equation for a torus: $v - e + t = 0$. (A derivation of this equation appears in the next Note.) We also use the fact that every edge connects a pair of vertices. Since there are $\binom{v}{2} = {}^{v(v-1)}\!/_2$ possible pairs of vertices, we have

$$e \le \frac{v(v-1)}{2}.$$

So $e = {}^3\!/_2 t$ and $v = e - t = {}^1\!/_2 t$ yields

$$3t \le \frac{t}{2}\left(\frac{t}{2} - 1\right),$$

from which it follows that $t \ge 14$.

Regarding a torus as a square with opposite edge identifications, we can demonstrate a triangulation that uses precisely 14 triangles.

Challenge. What does this triangulation look like on the fully formed torus?

14.5. Platonic Bagels

Let v be the number of vertices, e the number of edges, and r the number of regions in a pentagonal decomposition of a bagel. Since m edges meet at each vertex, $mv = 2e$ (each edge is counted twice). Also, since each region has 5 edges, $5r = 2e$ (again each edge is counted twice). Using these equations along with Euler's equation $v - e + r = 0$ (see section 14.4), we obtain $m = {}^{10}\!/_3$, which is impossible! No such "platonic bagel" exists.

Challenge. Do other platonic bagels exist? Is it possible to cover a bagel with triangles, squares, or even 13-gons in such a way that the same number of edges meet at each vertex?

Acknowledgments and Further Reading

Columbus, by sailing west and hoping to return from the east, demonstrated his belief that the earth was not flat. The modern world guessed it was a sphere but it could have turned out to be a torus! If we ever send a brave astronaut into space and many years later she returns from the opposite direction, we'll likely conclude that the universe is not flat. It could be a four-dimensional torus! To see what I mean by this (plus for many more bagel problems) have a look at [Tant1]; see also Jeff Week's fabulous book [Week].

The *Jordan Curve Theorem* states that any simple closed curve drawn on a plane or a sphere has the property of dividing what it is drawn on into two separate pieces. This "utterly obvious" theorem is surprisingly difficult to prove. Its validity says something significant about the geometry and structure of surfaces on which curves are drawn (it's not true for tori, for example). This geometry needs to be fully understood before one can properly attempt a proof. The brave reader can have a look at James Vick's book [Vick], chapter 1, for a proof of the Jordan Curve Theorem.

Euler's equation for spheres and multi-donuts (see the following Note) is presented a little more rigorously in B. H. Arnold's book [Arno]. More general formulas (including the case for Klein bottles, real projective planes, and the like) can be found in [Firb] and [Mass].

A Note on Euler's Equation

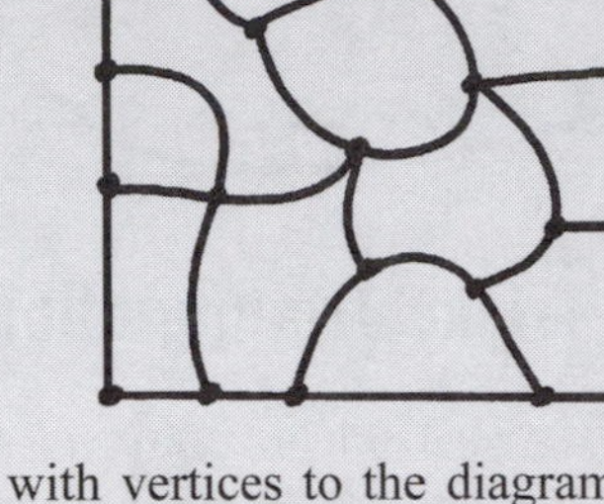

In the mid-1700s Leonhard Euler discovered a remarkable formula relating the number of regions, edges, and vertices for any diagram (or map) drawn on a plane or a globe (sphere). If r is the number of regions (including the outer boundary region in a planar map), e the number of edges, and v the number of vertices, then

$$v - e + r = 2.$$

In the diagram, for example, $v = 20$, $e = 30$, and $r = 12$. There is one proviso, however: We need to make sure that no region contains a "hole." Adding an extra edge with vertices to the diagram will always remove this difficulty.

→

The proof of Euler's formula is an induction argument on the number of edges. If there is just one edge, then the diagram must be of the necklace form shown here. For this case, $v = 1$, $e = 1$, and $r = 2$, so Euler's formula holds true.

Suppose the formula is true for diagrams involving k edges. Consider a diagram with $e = k + 1$ edges, r regions, and v vertices. Modify it by removing an edge that separates two distinct regions and melding those two regions together. (Why does such an edge always exist? Would you still call a diagram with no such edge a "map"? See [Char].) This produces a new map with $e - 1 = k$ edges, $r - 1$ regions and v vertices. By the induction hypothesis, Euler's formula holds true for this new diagram:

$$v - (e - 1) + (r - 1) = 2.$$

It follows that $v - e + r = 2$ holds for the original diagram. This completes the proof by induction.

For diagrams on a torus we have a different formula,

$$v - e + r = 0,$$

provided again that no region contains a hole or bounds a loop. (Add extra edges if necessary.) The proof of this formula actually follows from that for a sphere. Given a map on a torus, draw a circle that wraps around the tube of the torus cutting through edges and regions

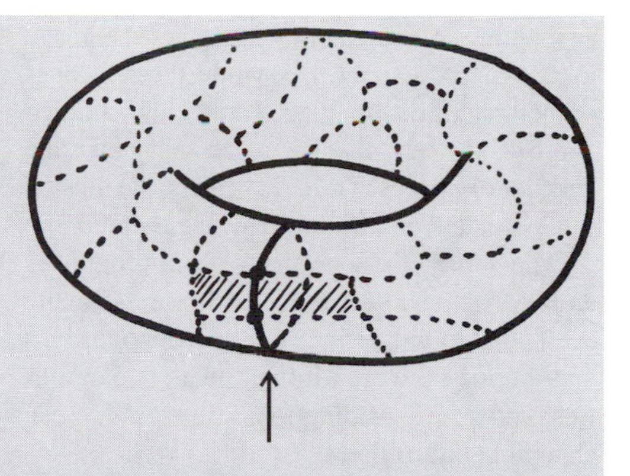

of the diagram in the process. (Just make sure this circle never follows an entire edge or a portion of an edge in the diagram, but instead cuts through edges at single points.) This circle introduces new edges and vertices and divides some regions of the diagram in two. Suppose n vertices are introduced. Each vertex cuts an existing edge in two and so produces n additional edges. Also, between pairs of vertices there are n new edges slicing n regions in two. Thus we have introduced a total of n new vertices, $n + n = 2n$ new edges, and n regions into the diagram. Notice, however, that the value of $v - e + r$ is unchanged!

Now cut the torus along this circle as shown in the figure below. This doubles the number of edges and vertices along the circle and produces two new regions. The value of $v - e$ has not changed, but the value of r has now increased by 2. But this new object is really a diagram on a (deformed) sphere, for which we know $v - e + (r + 2) = 2$. Thus it follows that for the original diagram on the torus we must have had $v - e + r = 0$. This proves Euler's formula for a torus.

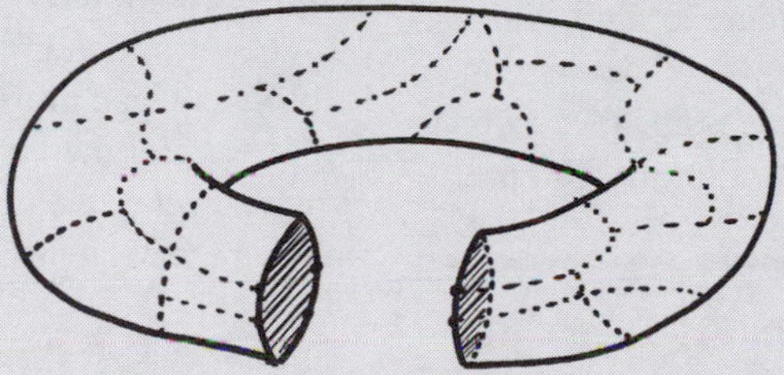

Challenge. A diagram of vertices, edges and regions is drawn on a *multi-donut* with p holes. Prove that $v - e + r = 2 - 2p$.

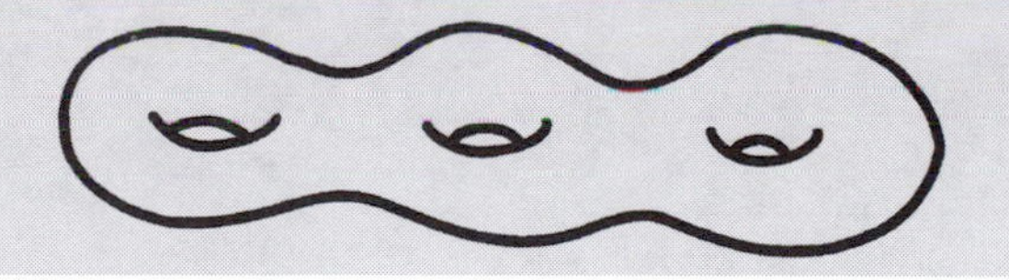

15 Capturing Chaos

15.1 Feedback Frenzy

A video camera pointed at its own screen processes that image over and over again. But in our set-up we have introduced some blatant errors—the image is at a tilt and the magnification is not exactly one-to-one (unless you have adjusted the zoom so that precisely the true sizes of objects are displayed). In just a few seconds a video camera processes its image hundreds of times, so very quickly the effects of these errors magnify. The cumulative effect can be stunningly beautiful. (Try poking your finger into the image. What happens?)

There is a slight delay in the circuitry between the instant the camera sees a change in the status of an object (such as the extinguishing of a flame) and the moment this change appears on the screen. It takes twice this time for the image of the image to change, three times as long for the image of the image of the image to change, and so on. All the while, these changes in the images of the images are being recorded and causing new changes to be sent down the corridor of images. The effect can be the illusion of a persistent and beautiful swirling motion.

The same effect occurs with microphones and PA systems. The amplifier transmits minute erroneous sounds, which the microphone hears and amplifies, which it then hears and amplifies, and so on, leading to a situation that can easily get out of hand!

15.2 Creeping up on Chaos

Somewhere around $r = 2.92$ the sequence bifurcates. Between $r = 3.4$ and $r = 3.5$ the *period* of the oscillation *doubles* again and the sequence begins to oscillate among four values rather than two. The period doubles again near $r = 3.545$ to eight values and then again to sixteen around $r = 3.565$. Plotting 100 or 200

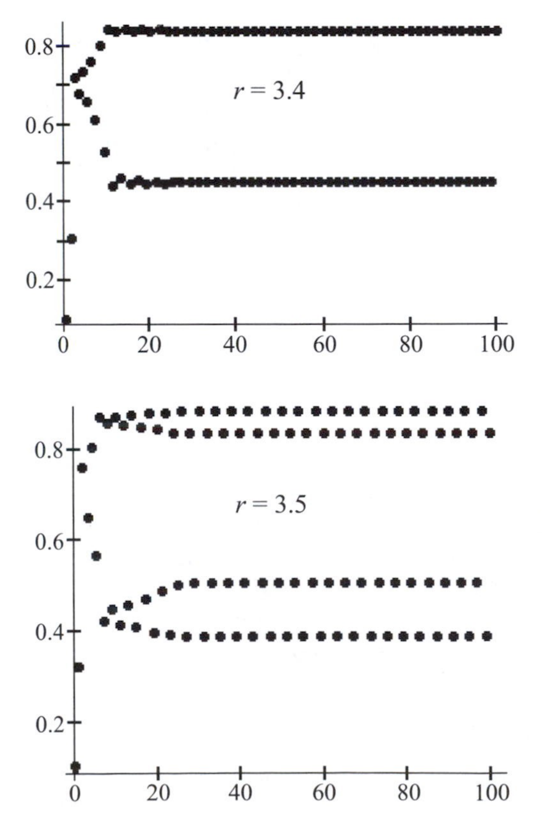

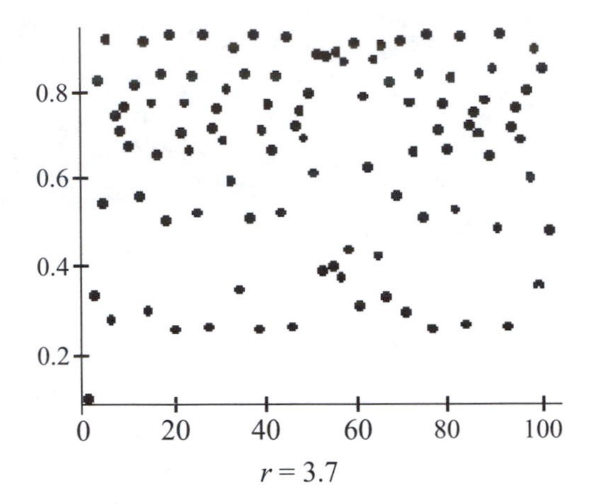

terms helps make this clear. With higher values of r it is difficult to discern the oscillations among the larger numbers of terms, and certainly by $r = 3.7$ the behavior of the sequence values is completely unpredictable.

The introduction of period doubling and the subsequent lead into chaos is all the more striking if we plot the limit values of the sequence with respect to the value of r. For example, the limit of the sequence is 0.5 when $r = 2$.

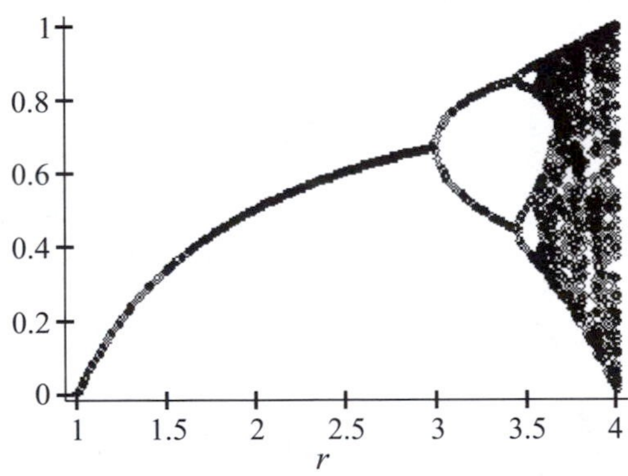

It is remarkable that this simple recursive equation leads to such complex behavior! This peculiarity was first observed by the mathematical biologist Robert May in his 1974 study of insect population levels. It was one of the first examples of chaos in mathematical systems.

Challenge 1. In this section we worked with the initial value $a_0 = 0.1$. Choose a different value between 0 and 1. Do the results you obtained above change? How *sensitive* is this experiment to initial conditions? See [Hall], chapter 3.

Challenge 2. We never saw period *tripling* occur. Why not? *Hint:* Our equation is quadratic.

A Note on Population Models

(*Warning:* Calculus!) Population models, be they to predict the world's human population in the year 2363 or the biomass of a yeast culture by the day's end, often are based on the assumption that the size of the population varies continuously with time. This is often a reasonable approximation. Given the current size of the world's population, for example, someone is born or dies at almost every instant. Thus if $P(t)$ represents the size of a population at time t, it is convenient to assume that P is a continuous function of time. To bring in the power of calculus, we assume moreover that P represents a *differentiable* function of time; after all, the derivative of a function indicates its rate of change, a concept physically significant in population growth.

A popular, basic population model states simply that the rate of change of a population at any instant is proportional to the population at that instant:

$$\text{Rate of Change} = {}^{dP}/_{dt} = rP \quad \text{for some constant } r.$$

For instance, according to the Population Reference Bureau, the current growth rate of the human population is 1.4% per year, thus we have ${}^{dP}/_{dt} = 0.014P$. Solving the rate-of-change equation yields

$$P(t) = P(0)e^{rt}.$$

This model may be helpful for short-term predictions but it cannot be relied upon for long-term analysis. The exponential function grows without bound, but we know that population sizes cannot: all population environments have a certain *carrying capacity*. Earth, for example, has only a finite supply of resources and cannot sustain an arbitrarily large population size.

To rectify this problem, we modify the simple population model and insert a term to reflect a maximal carrying capacity of a population of size M. Consider the formula

$$\text{Rate of Change} = {}^{dP}/_{dt} = rP(M - P).$$

Notice that if the population size P is below the carrying capacity value, $0 < P < M$, then ${}^{dP}/_{dt} > 0$ meaning that population levels will grow (increase). If the population size is ever larger than M, then ${}^{dP}/_{dt} < 0$ and the population levels will fall—just as we expect to happen. This model is called the *Logistic Growth Model*.

It is possible to solve this differential equation and write down an explicit formula for $P(t)$ (using separation of variables), but we can easily get an accurate picture of its graph without all this work. As we have already noted, ${}^{dP}/_{dt} > 0$ for $P \in (0, M)$, so the function is increasing in this range. If $P > M$, it is decreasing. Differentiating once through the logistic growth equation yields

$$P'' = rP'(M - P) - rPP' = rP'(M - 2P),$$

from which it follows that the curve is concave up ($P'' > 0$) for $0 < P < {}^{M}/_{2}$ and for $P > M$, and concave down ($P'' < 0$) for ${}^{M}/_{2} < P < M$. Depending on the value of the initial population, the logistic growth curve thus has one of three forms illustrated here.

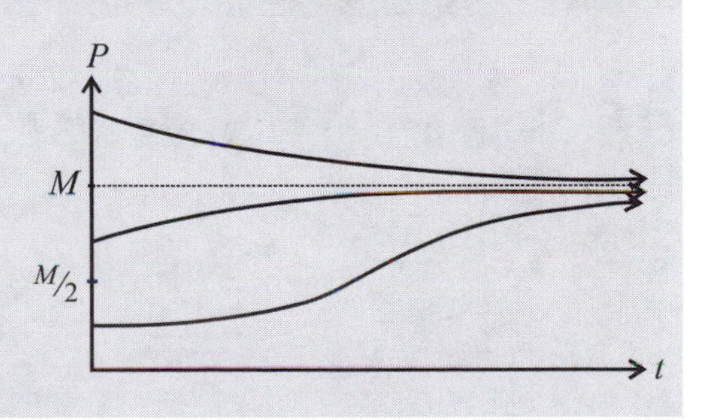

Many populations do in fact follow this model reasonably well; for example, [Gior]

discusses modelling the growth of a yeast culture with this equation. But other populations, such as some insect populations do not. Could it be because some insects live only a few summer weeks or months their population levels change at discrete times from year to year? Is the continuous model appropriate?

Let's now suppose population levels are recorded at discrete units of time: once a year. Let P_n represent the population in the nth year. The appropriate modification of the logistic growth equation would be:

$$\text{Rate of change} = \frac{P_{n+1} - P_n}{(n+1) - n} = rP_n(M - P_n).$$

That is

$$P_{n+1} = rP_n(M - P_n) + P_n = rP_n\left[\left(M + \frac{1}{r}\right) - P_n\right].$$

Up to modification of the constants involved, this is the recursive equation presented in section 15.2. Given the solution to the *continuous* logistic equation, we would expect the population levels to behave in a reasonably predictable way with time and perhaps follow the same shape when graphed as the logistic curve. However, as we saw in 15.2, this is not the case! Depending on the value of r, the *discrete* logistic growth model can represent purely chaotic population values! As Robert May noted, population levels of some species do in fact vary chaotically from year to year.

This leaves a delicate question: Since the continuous model works so well in some population studies and not in others, how do we decide which model a species' population is likely to follow?

Acknowledgments and Further Reading

There are many wonderful introductory books for nonlinear dynamics and chaos. In the beautiful and accessible collection of essays [Hall] edited by Nina Hall, Sir Robert May writes about his work in chapter 7. Also check out [Deva2] for an elementary discussion on period doubling as well as a study of fractals. Population modelling is beautifully discussed in John MacQueen's essay "Biological Modelling: Population Dynamics" [Bond], chapter 7.

16 Who has the Advantage?

16.1 A Fair Game?

The game is absolutely fair! Irrespective of the number of coins, as long as one player has exactly one more than the other, each participant is equally likely to win. Suppose player one has $n + 1$ coins and player two n coins. If in a game player one receives H_1 heads and T_1 tails, then necessarily $H_1 + T_1 = n + 1$. If player two receives H_2 heads and T_2 tails then $H_2 + T_2 = n$. Player one wins if $H_1 > H_2$.

Note that both $H_1 \le H_2$ and $T_1 \le T_2$ cannot hold, for adding these inequalities would yield the absurdity $n + 1 \le n$. Thus either $H_1 > H_2$ or $T_1 > T_2$ holds. Further, both cannot hold at the same time, for adding the inequalities $H_1 \ge H_2 + 1$ and $T_1 \ge T_2 + 1$ yields another absurdity, $n + 1 \ge n + 2$.

Thus exactly one of $H_1 > H_2$ and $T_1 > T_2$ holds. But both are equally likely (the game could just as well have been phrased in terms of tails rather than heads), so

$$P(H_1 > H_2) = ½ = P(T_1 > T_2).$$

Thus player one has a 50% chance of winning, and consequently so too does player two.

Challenge. If one player has *two* more coins than the other, what now are each player's chances of winning?

16.2 Voting for Pizza

Here is a table illustrating the voting options available to Brad (A > O > P) and Cassandra (O > P > A) and the resultant outcomes, given it is known that Alice will vote pepperoni.

		Brad		
		P	A	O
Cassandra	P	P	P	P
	A	P	A	P
	O	P	P	O

A = anchovies, O = olives, P = pepperoni

Looking at the table, Brad realizes that Cassandra will not choose A (second row). Any other vote will guarantee her an outcome better than A. Brad therefore will not choose P or A (first and second columns), for otherwise he will be landed with his least favorite topping, P. At least by choosing O, there is a chance he will do better.

Cassandra, by examining the table, knows Brad will reason this way and choose O. Her move is thus to vote O and be assured of a pizza with her favorite topping!

The tie-breaking advantage given to Alice was by no means an advantage! In fact it was positively harmful and guaranteed Alice her least liked pizza topping. If Alice had known her friends were savvy and would reason in this manner, perhaps she would have lied during their initial discussions on topping preferences!

Challenge. Imagine a situation where the three friends again share their preference lists before taking a vote. This time Alice lies and announces that A is her favored choice (with P and O ranking in some order below). What would happen if Alice then indeed wrote down A in a vote? What would happen if, despite her announcement, she actually wrote down P, her secretly favored choice?

16.3 A Three Way Duel

Surprisingly, deliberately missing his first shot dramatically increased Alberto's chances of survival from about 31% to about 40%! (Bridget's chances of survival decreased to about 38% but Case's increased to 22%). Letting the two "superpowers" battle it out between themselves before entering the action turns out to be the smartest move!

The following Note shows how to derive these probabilities mathematically.

A Note on Computing Probabilities in a Three-Way Duel

We assume that Case, denoted C, and Bridget, B, always follow their strategy of aiming for the most competent participant standing. To begin with, let's assume Alberto, A, also follows this plan, even on his first shot. The tree diagram on the next page displays all possible outcomes of the duel. The nodes indicate who is still standing at that time of play. Note that there are scenarios of arbitrary length. We are assuming that all participants have an infinite supply of bullets and infinite stamina for this "truel!"

Let's read off Alberto's chance of survival. It comes from following all the possible paths that lead to a single standing A, multiplying all the probabilities encountered along the way:

$$\begin{aligned} P(A \text{ wins}) &= \left(\tfrac{1}{3}\cdot\tfrac{1}{3}\cdot\tfrac{1}{3}+\tfrac{1}{3}\cdot\tfrac{1}{3}\cdot\tfrac{2}{3}\cdot\tfrac{1}{3}\cdot\tfrac{1}{3}+\tfrac{1}{3}\cdot\tfrac{1}{3}\cdot\tfrac{2}{3}\cdot\tfrac{1}{3}\cdot\tfrac{2}{3}\cdot\tfrac{1}{3}\cdot\tfrac{1}{3}+\cdots\right) \\ &\quad +\left(\tfrac{2}{3}\cdot\tfrac{2}{3}\cdot\tfrac{1}{3}+\tfrac{2}{3}\cdot\tfrac{2}{3}\cdot\tfrac{2}{3}\cdot\tfrac{1}{3}\cdot\tfrac{1}{3}+\tfrac{2}{3}\cdot\tfrac{2}{3}\cdot\tfrac{2}{3}\cdot\tfrac{1}{3}\cdot\tfrac{2}{3}\cdot\tfrac{1}{3}\cdot\tfrac{1}{3}+\cdots\right)+\tfrac{2}{3}\cdot\tfrac{1}{3}\cdot\tfrac{1}{3} \\ &= \tfrac{1}{27}\left[1+\tfrac{2}{9}+\left(\tfrac{2}{9}\right)^2+\cdots\right]+\tfrac{4}{27}\left[1+\tfrac{2}{9}+\left(\tfrac{2}{9}\right)^2+\cdots\right]+\tfrac{2}{27}. \end{aligned}$$

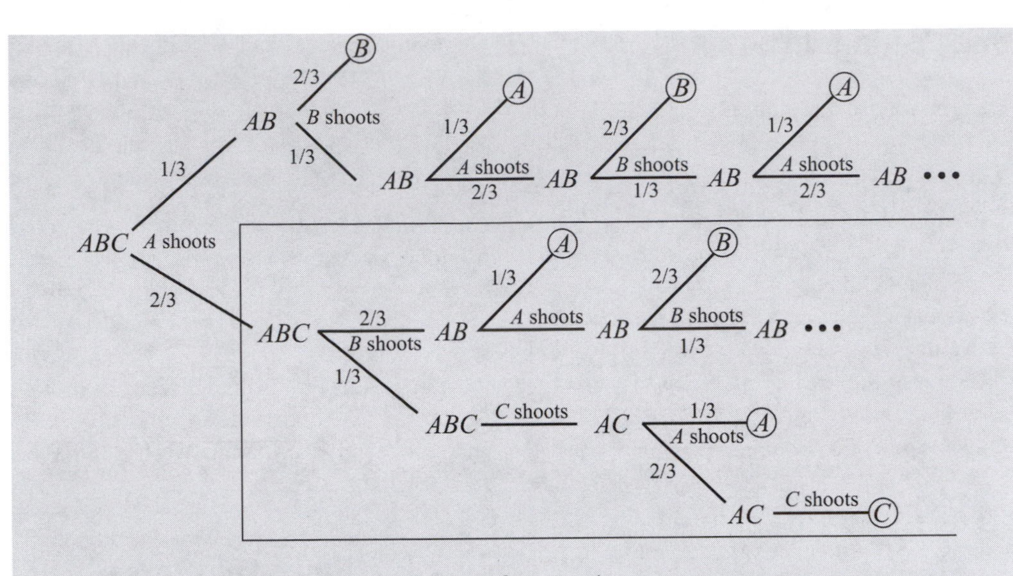

Using the geometric series formula $1 + x + x^2 + \cdots = 1/_{1-x}$ for $0 < x < 1$, we have

$$P(A \text{ wins}) = \frac{1}{27} \cdot \frac{1}{1-\frac{2}{9}} + \frac{4}{27} \cdot \frac{1}{1-\frac{2}{9}} + \frac{2}{27} = \frac{59}{189} \approx 31\%$$

In the same way we can compute $P(B \text{ wins}) = 34/63 \approx 54\%$ and $P(C \text{ wins}) = 2/3 \cdot 1/3 \cdot 2/3 = 4/27 \approx 15\%$.

Suppose now that Alberto deliberately misses his first shot. The tree diagram of all possible outcomes is the boxed region in the tree above. Thus Alberto's chances of survival are now

$$P(A \text{ wins}) = \left(\frac{2}{3} \cdot \frac{1}{3} + \frac{2}{3} \cdot \frac{2}{3} \cdot \frac{1}{3} \cdot \frac{1}{3} + \frac{2}{3} \cdot \frac{2}{3} \cdot \frac{1}{3} \cdot \frac{2}{3} \cdot \frac{1}{3} \cdot \frac{1}{3} + \cdots \right) + \frac{1}{3} \cdot \frac{1}{3}$$

$$= \frac{2}{9}\left[1 + \frac{2}{9} + \left(\frac{2}{9}\right)^2 + \cdots \right] + \frac{1}{9} = \frac{2}{9} \cdot \frac{1}{1-\frac{2}{9}} + \frac{1}{9} = \frac{25}{63} \approx 40\%.$$

Similarly $P(B \text{ wins}) = 8/21 \approx 38\%$ and $P(C \text{ wins}) = 2/9 \approx 22\%$.

Challenge 1. Suppose instead, on average, Alberto successfully hits a target one quarter of the time, Bridget one third of the time, and again Case is a perfect shot. In this game, is it still better for Alberto to deliberately miss his first shot?

Challenge 2. Is there a scenario, with Alberto still the weakest player and Case a perfect shooter, in which deliberately missing his first shot is *not* beneficial to Alberto?

Challenge 3. Suppose Alberto, Bridget, and Case are all perfect shots. Playing the same game, what strategy would each player adopt? What will result?

16.4 Weird Dice

Die A beats die B two thirds of the time. It is also easy to see that B beats C two thirds of the time too. C beats D if C shows a 2 and D a 1 or C shows a 6. Thus

$$P(\text{C beats D}) = \frac{2}{3} \cdot \frac{1}{2} + \frac{1}{3} = \frac{2}{3}.$$

Finally, D beats A if D shows a 1 and C a 0 or if D shows a 5. Thus

$$P(\text{D beats A}) = \frac{1}{2}\cdot\frac{1}{3}+\frac{1}{2}=\frac{2}{3}.$$

Do not play this game with me. I will win two thirds of the time simply by choosing the die preceding whichever one you chose (or die D if you chose die A)!

These dice exhibit surprising nontransitive behavior. If we are told, for example, that Agatha is taller than Beatrice, and Beatrice is taller than Chanice, then, without ever having met these folk, we can be certain that Agatha is taller than Chanice. The relation "taller than" is certainly transitive. But this need not always be the case for all systems. Here we have four dice with the property that A usually beats B, B usually beats C, C usually beats D, and D usually beats A—all with the same probabilities!

Notice that we can throw out die D and be left with three dice with this same nontransitive behavior (but not the same probability values throughout: C beats A five ninths of the time.)

Challenge 1. Three dice A, B, and C have the property that A beats B, B beats C, and C beats A all with the same probability. Can you find them? See [Berl2].

Challenge 2. Do there exist three (or perhaps necessarily more) tetrahedral, octahedral, dodecahedral, or isocahedral dice with this property?

A Note on Social Choice Theory

Nontransitive behavior also occurs in the theory of *social choice*— the theory of voting and election procedures. We have already seen an occurrence of this in Section 16.2. Recall that Alice preferred a pepperoni pizza topping to anchovies to olives, Brad preferred anchovies to olives to pepperoni, and Cassandra olives to pepperoni to anchovies. If these students decided on a simple "majority rules" voting scheme, then pepperoni beats anchovies, anchovies beats olives and olives beats pepperoni each by a 2:1 vote! This phenomenon is called the *Condorcet paradox,* a phenomenon believed to have been first observed by the Marquis de Condorcet (1743–1794). Many bizarre paradoxes occur in social choice theory.

Consider four major voting schemes in operation today. Each involves the voters ranking the candidates in order of preference. The winner of the election is determined as follows:

1. *Plurality Voting.* The candidate receiving the largest number of top choice nominations wins. (This is equivalent to a simple "majority rules" scheme where each voter submits just a single name on a ballot.)

2. *The Borda Count.* Each candidate receives points for her position on a voter's preference list: n points (if there are n candidates) for each first-place ranking down to 1 point for each last-place ranking. The candidate accruing the most points overall wins. (This scheme is often used to rank U.S. football teams.)

3. *Sequential Pairwise Counting.* Two candidates are pitted against each other in a head-to-head match. The winner is then pitted against a third candidate, and so on down a list of all the candidates until there is a single winner. (Similarly, sports teams often play elimination rounds until there is a single winning team.)

4. *The Hare System.* The candidate receiving the fewest first-place rankings is eliminated. Amongst the remaining candidates, the one receiving the fewest first-place rankings in the modified preference lists is eliminated, and so on, until a single winner remains.

Each of these voting systems is fundamentally flawed as the following example demonstrates. Suppose a professor offers to bring pizza to math club. He allows the students to vote on one of four possible toppings: anchovies, pepperoni, olives, or sausage.

1. *Plurality Voting.* Suppose the seven students in the club vote as indicated in this table.

Choice	1	1	1	1	1	1	1
First	A	A	P	P	A	O	S
Second	S	S	S	S	P	P	P
Third	O	O	A	A	S	S	A
Fourth	P	P	O	O	O	A	O

Under plurality voting anchovies wins with three votes (look at the top choices.) However, there is something curious about the status of pepperoni. If it were a vote just between pepperoni and anchovies, pepperoni would have won (by a 4:3 vote). The same is true if pepperoni is matched against any other single option! (All 4:3 wins!) Pepperoni beat every other alternative one-on-one, but did not win the election as a whole. This is odd and many people deem this phenomenon a serious flaw of Plurality voting. ■

2. *The Borda Count.* The professor is interested only in the students' opinions about pepperoni and olives. When polled, the students respond as the table indicates.

Choice	1	1	1	1	1	1	1	Points
First	P	P	P	P	O	O	O	2
Second	O	O	O	O	P	P	P	1

According to the Borda count, pepperoni receives $4 \times 2 + 3 \times 1 = 11$ points so is preferred over olives with 10 points.
The students' opinions about anchovies are irrelevant to the professor's concerns but he decides to ask about them in any case. This time he tabulates the responses as shown.

Choice	1	1	1	1	1	1	1	Points
First	P	P	P	P	O	O	O	3
Second	O	O	O	O	A	A	A	2
Third	A	A	A	A	P	P	P	1

Four students still indicate that they prefer pepperoni to olives and three prefer olives to pepperoni. However, according to the Borda count, the group overall now prefers olives (with 17 points) to pepperoni (15 points)! Asking for irrelevant information has altered the conclusion! This is deemed a serious flaw with the Borda count. ■

3. *Sequential Pairwise Voting.* This time just three students are asked to rank anchovies, pepperoni, olives, and sausage. The table indicates their preferences.

Choice	1	1	1
First	P	A	O
Second	O	P	S
Third	S	O	A
Fourth	A	S	P

Let's first pit pepperoni against olives, then the winner against anchovies, and the second winner against sausage.

Round 1: Pepperoni vs. Olives. Pepperoni wins with two votes.

Round 2: Pepperoni vs. Anchovies. Anchovies wins with two votes.

Round 3: Anchovies vs. Sausage. Sausage wins with two votes.

This leaves sausage as the overall winner. But notice something curious: Everyone preferred olives to sausages! How did sausages end up winning the election? Many people deem this phenomenon a serious flaw of sequential pairwise voting. ■

4. *The Hare System.* Nineteen students vote for anchovies, pepperoni, olives, and sausage as shown in the table.

Choice	7	6	4	2
First	P	A	O	S
Second	S	P	A	O
Third	O	O	S	P
Fourth	A	S	P	A

Round 1. Receiving only two top-choice votes, sausage is eliminated first.

Round 2. With sausage gone, pepperoni receives seven top-choice votes, anchovies six, and olives $4 + 2 = 6$ top-choice votes. Both anchovies and olives are eliminated, leaving pepperoni as the winner.

However, the professor is unsure of some handwriting and decides to redo the election. Suppose everyone votes the same way, except the two students in the far right column who each bump pepperoni up one position on their preference lists. The new vote is shown in the table, with the changes in boldface.

Choice	7	6	4	2
First	P	A	O	S
Second	S	P	A	**P**
Third	O	O	S	**O**
Fourth	A	S	P	A

Pepperoni won the first election. Surely moving pepperoni higher on a preference list will ensure pepperoni's success a second time. However, this is not the case. This change, supposedly in favor of pepperoni, causes anchovies to win! (Check it!) This phenomenon is deemed a serious flaw of the Hare system. ■

People have been searching for centuries for a 'perfect' voting procedure, one devoid of paradoxical behavior and philosophical flaws, but always failed to find one. The search ended when, in 1951, Kenneth Arrow, an economist at Harvard University, proved a deep and disturbing result:

Arrow's Theorem. *No democratic voting procedure avoids all four flaws mentioned above.*

According to Arrow, any voting procedure a democratic society devises, no matter how ingenious or complex, is doomed to violate at least one of four basic requirements.

So where does this leave us today? Often, practical concerns dictate the type of social choice mechanism adopted. In the election of a student body president among six candidates, say, in a college of 1800 students, it would be infeasible to review individual preference lists. A simple one-name-per-ballot plurality procedure may be simplest and sufficient. Usually, practical common sense wins out in the end. Of course, if there is contention about which electoral procedure is best, a society can always take a vote on which one to use!

Acknowledgments and Further Reading

Arrow's Theorem is discussed thoroughly in Paul Hoffman's wonderful book [Hoff] and briefly by Ravi Vakil [Vaki]. An excellent elementary overview of voting theory appears in [COMA], chapter 11, which gives the real world example of the 1980 U.S. Senate race in New York where the paradox of plurality voting actually did occur. D. Saari and F. Valognes' article [Saar] is well worth a look.

The truel is a famous problem in probability theory that is still being studied today [Kilg], [Knut]. The nontransitive dice also appear in [Berl2], [Hons3], and [Vaki], and the coin game in [Vaki].

17 Laundry Math

17.1 Turning Clothes Inside Out

Taking it Further Answer. At first this question seems alarming. We could argue that a closed sphere has an interior space and an exterior space, and that the interior will always remain interior and the exterior space exterior no matter how we stretch, bend, or pull the surface. Everting the sphere would somehow mean switching the two spaces, which, quite simply, is impossible—if we stick with the current rules of the game. Imagine, however, a super high-tech material with the amazing ability to pass through itself. Self-intersection makes the eversion of a closed sphere possible. All one need do is push the north pole of the sphere down through the south and the south pole up through the north!

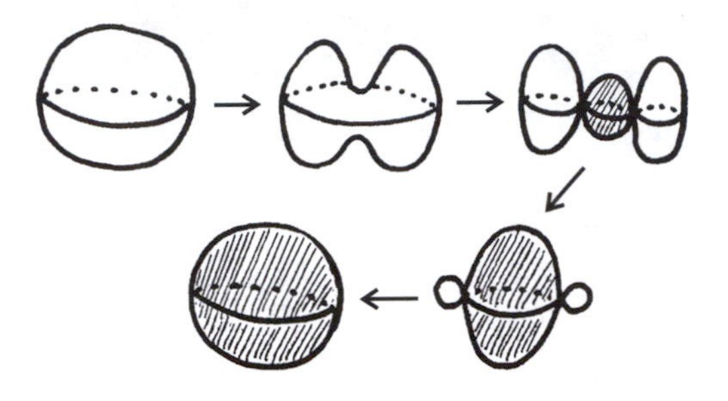

On one level this is a satisfactory solution, but on another it is not. Notice that a crease forms around the equator of the sphere as the sphere is pushed through itself. As we keep pulling, this crease will become sharp and the material is likely to snap! Is there a way to perform this eversion so as to avoid all sharp creases and consequent damage to the material?

In 1957 Steve Smale published his amazing discovery that a *smooth* eversion of a sphere is indeed possible. His celebrated work was a magnificent piece of theory but had one practical drawback: It didn't actually show how to perform the eversion. It took another seven years before Arnold Shapiro found a practical method of eversion, but even his technique was hard to comprehend. It wasn't until 1974 when Bill Thurston developed a scheme of "corrugations" that a comprehensible method was finally demonstrated.

The Geometry Center at the University of Minnesota has produced a computer-animated video, *Outside In* [Levy] that lets you see this eversion in all its splendor. It is a magnificent video well worth investigating. After watching you might wonder whether a smooth eversion of a non-punctured donut is possible using the same technique. (*Answer:* It is!)

Challenge. You could also evert a sphere by making use of the fourth dimension (See "A Note on the Fourth Dimension" starting on p. 135). Can you see how?

17.2 Mutilated Laundry

Even eversions of multi-donuts yields the same surfaces back again!

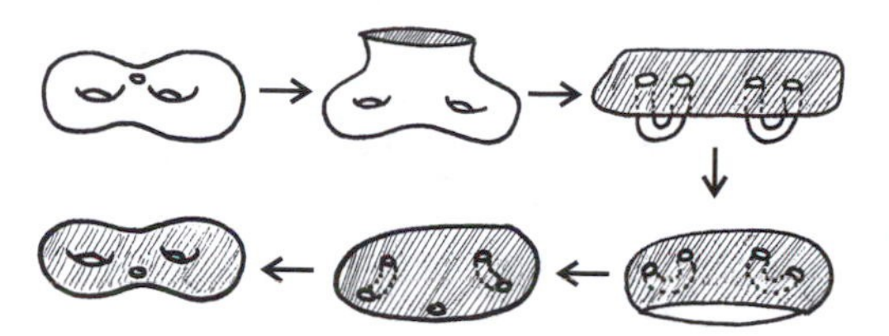

Taking it Further Answer. It turns out that the two alternative eversions of a donut we mentioned are physically distinct. No manipulation whatsoever (barring cutting and ripping) will convert one eversion into the other. To see why, imagine two rubber bands wrapped around the tube of the donut, one on the inside of the tube aligned in one direction, and the other on the outside of the torus in an orthogonal direction.

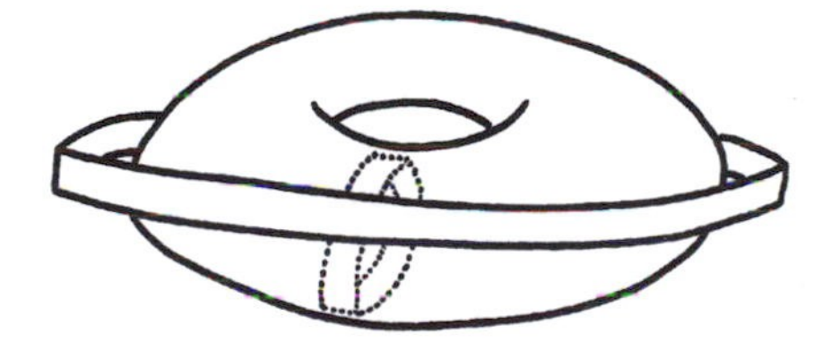

These two bands are unlinked and must remain so no matter how we manipulate the donut, even if we evert it. Our first type of eversion places the two bands into the following configuration.

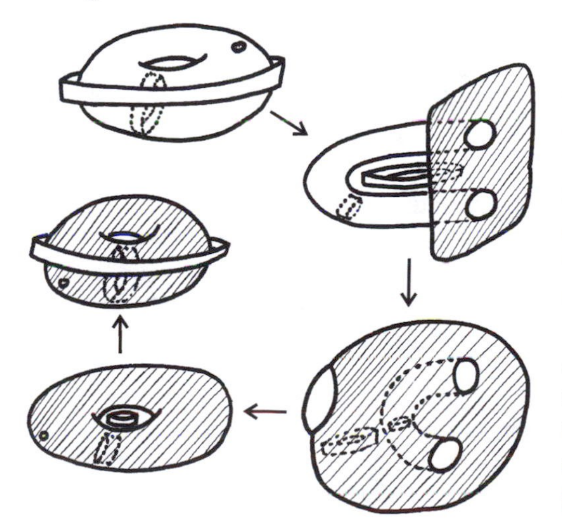

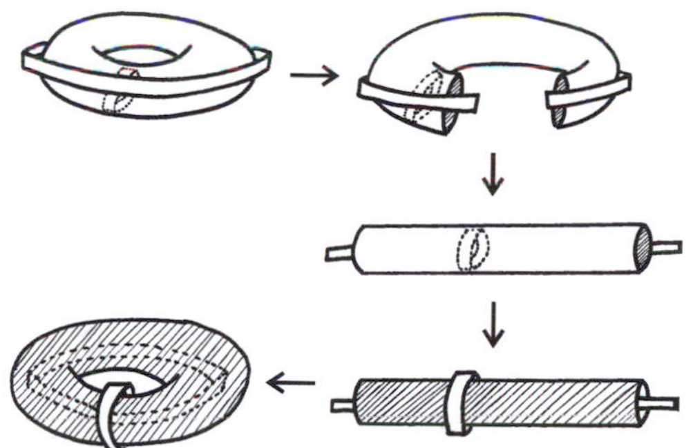

The second eversion (which, for now, I know how to perform only via cutting, everting, and repasting) produces the configuration above.

The two bands are now linked! As it is impossible to link two unlinked rings in 3-space, it must be impossible to convert our first eversion of the donut into the second via physical manipulations (avoiding cutting and tearing). The two eversions are distinct.

Challenge 1. Show by going to the fourth dimension that it is possible to manipulate one eversion of a punctured donut into the other (see section 10.2).

Challenge 2. How many physically distinct eversions of a punctured double-donut exist?

Challenge 3. Spheres, donuts, and multi-donuts are said to be *orientable* surfaces; they clearly have an "inside" and an "outside." However, there is another class of surfaces that I have side-stepped thus far.

Let's again start with a pair of trousers, but do not immediately sew the two leg openings together. First bring one leg up, over, and through the waist; then push it down through the tube of the second leg; now sew! The resulting shape is complicated, but it is one that the mathematical community recognizes. It is a punctured Klein bottle. (See also chapter 10.)

This is an example of a "one-sided" or *non-orientable* surface. An ant crawling on this shape could reach either side of the material without ever cheating by crossing over the edge of the hole. (Since we sewed the two legs together, the ant is allowed to cross over the new

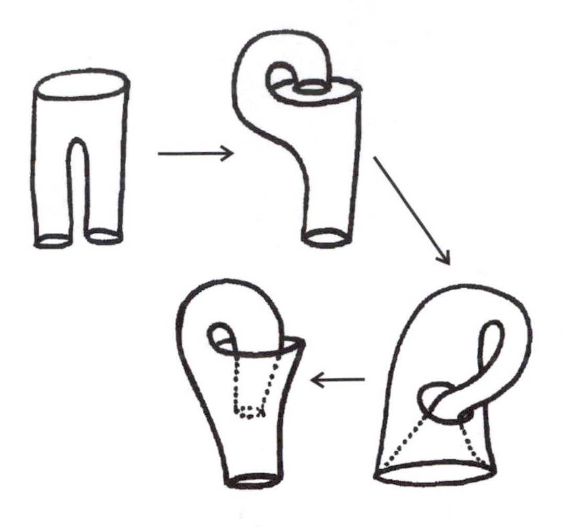

seamwork.) This surface has no "inside" or "outside" and the question of turning it inside out may be perturbing. Nonetheless, one can pull the material through itself to reveal an everted shape, and again it is exactly the same in structure. The eversion of a Klein bottle is another Klein bottle. (Try it!) The trouser material is, as expected, inside out, but this time the second leg is down inside the tube of the first.

You could also obtain an everted Klein bottle by first turning the pair of trousers inside out and then sewing the two leg openings together. Is this a new eversion of the Klein bottle, physically distinct from the first? Or is it just the first eversion in another guise?

17.3 Cannibalistic Clothing

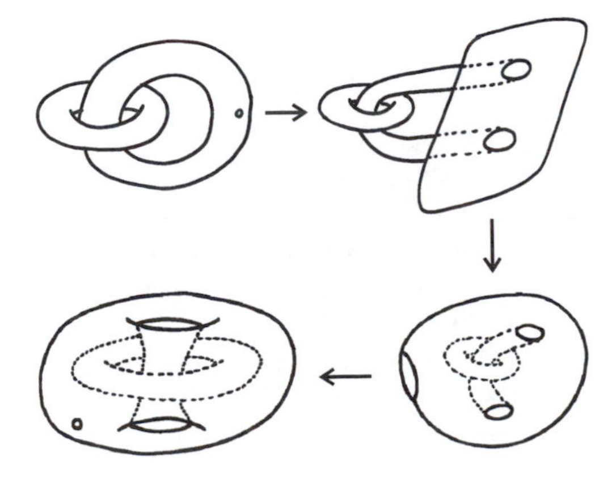

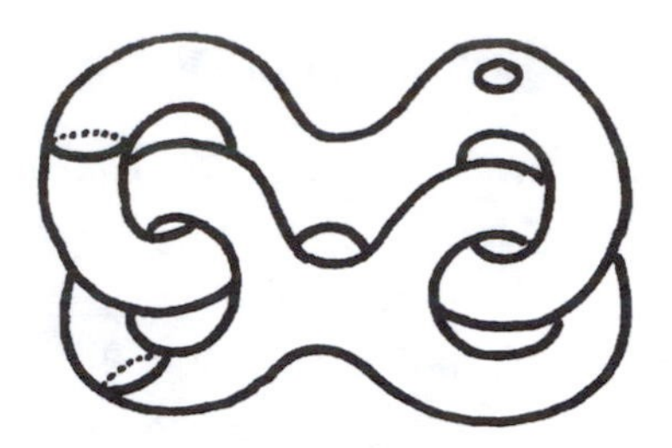

Challenge. What about carnivorous jumpsuits? Is it possible for a hungry double-donut to consume an unlucky peer? How would the victim sit inside the victor?

Acknowledgments and Further Reading

To learn more about Arnold Shapiro's eversion of a sphere have a look at [Fran1] and [Phil]. Martin Gardner writes about donut eversion problems in [Gard15], chapter 5. Also have a look at Herbert Taylor's whimsical article [Tayl].

18 Get Knotted

18.1 Party Trick I: Two Linked Rings?

It is impossible to unlink two linked rings, and the puzzle appears unsolvable. However, we are not actually dealing with two linked rings: a little space exists between Jason's wrist and the loop of string around it. To escape, Paul should push a fold of his string through this space, have Jason push his hand through this fold, and then withdraw the fold. This will do the trick!

Comment. This trick reminds me of another puzzler. Holding the ends of a long piece of string, one end in each hand, is it possible to tie a knot in the string without ever letting go? The

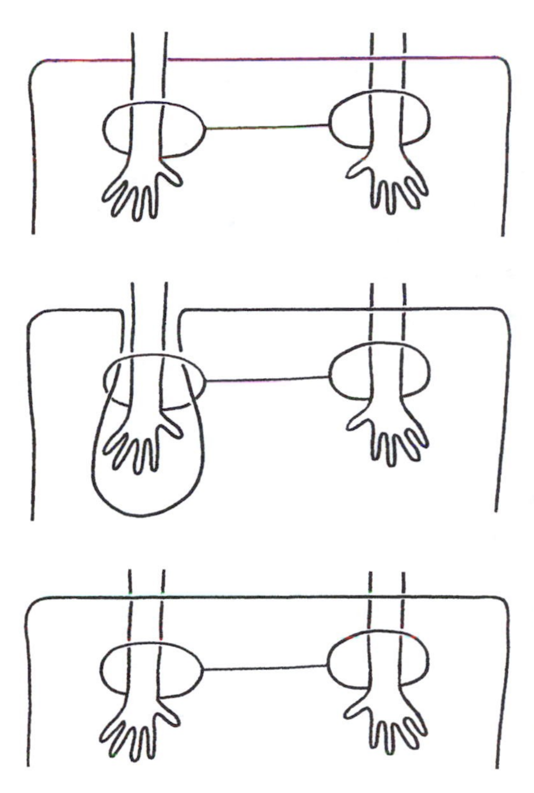

answer is yes—if you fold your arms first before taking hold of the string!

18.2 Party Trick II: A T-Shirt Trick

Take the T-shirt off George and let it dangle from his arms. Now push one sleeve back inside the shirt and, following his arms, pull it out through the other sleeve. This turns the shirt inside out and again leaves it dangling on his arms. Now put the shirt back over George's head and back on George.

Taking it Further Answer. Lifting the shirt straight off George and over on to Aliza does the trick.

Challenge 1. Is there a way to take the shirt off Aliza and put it back on George inside out?

Challenge 2. George and Aliza again hold hands but Aliza's arms are crossed over. What happens to the T-shirt when it is transferred from one person to the other? What if George and Aliza both cross their outstretched arms before holding hands? What if both people hold the ends of two styrofoam tubes twisted together multiple times? Is it possible to predict when a T-shirt transferred from one person to the other will arrive inside out and/or back-to-front?

Challenge 3. A prisoner is wearing handcuffs and his ankles are shackled together. He wants to turn both his shirt and his trousers inside out. Is this possible?

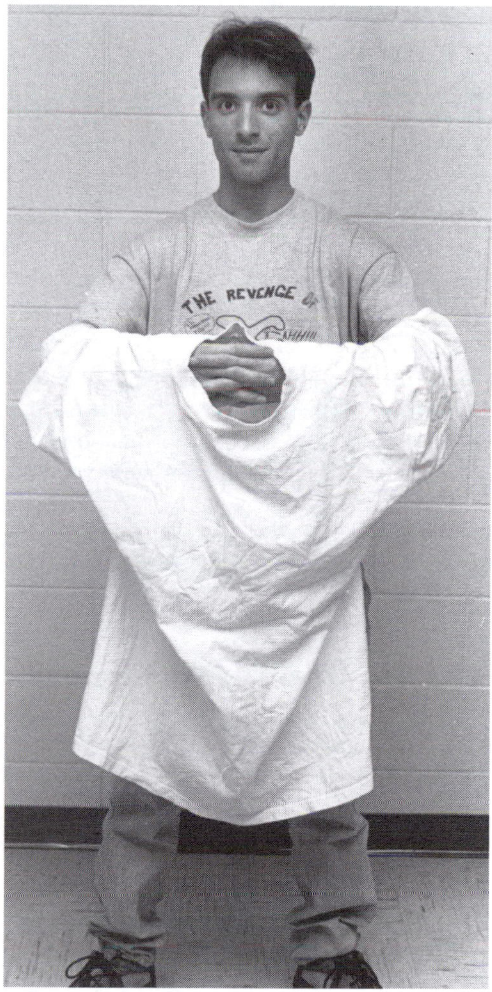

Have the shirt dangle from George's arms and then push one sleeve through the other.

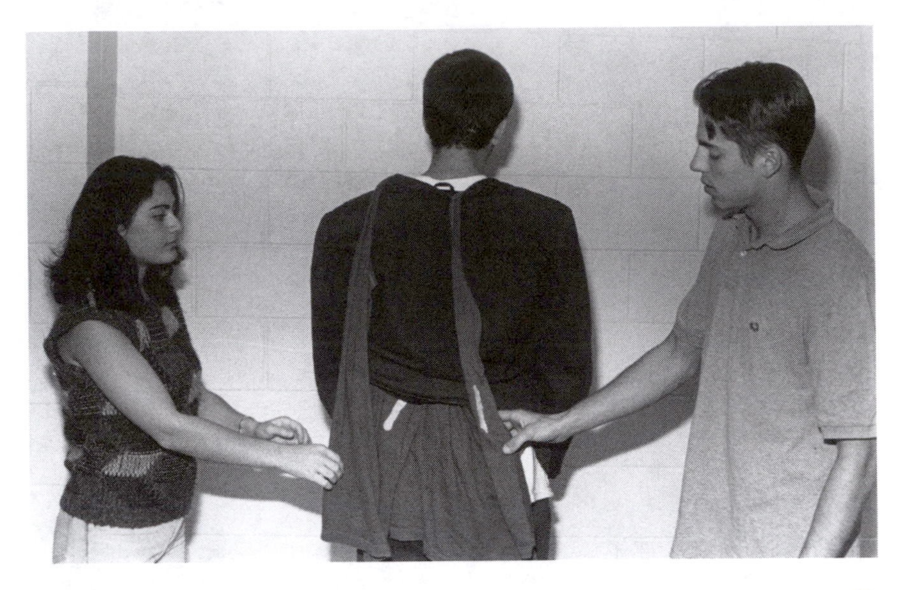

18.3 Party Trick III: A Waistcoat Trick

Slip both arms through the armholes of the waistcoat. This leaves the waistcoat dangling from the back material of the jacket much like the T-shirt dangling from George's arms in section 18.2. Using George's trick, turn the waistcoat inside out by pushing it through one of its armholes. Have Sten-Ove then place his arms back through the sleeves of the waistcoat and slip it back underneath his jacket. Voila! Trick complete!

Challenge. Is it possible to complete this trick if either the jacket or waistcoat (or both!) remains buttoned up?

18.4 Two More Linked Rings?

Continue to mold the stem of the loop though the second loop and then back to its original position.

Challenge. Show that these two Play-Doh sculptures are "equivalent," that is, you can convert one into the other by "smooth" manipulations.

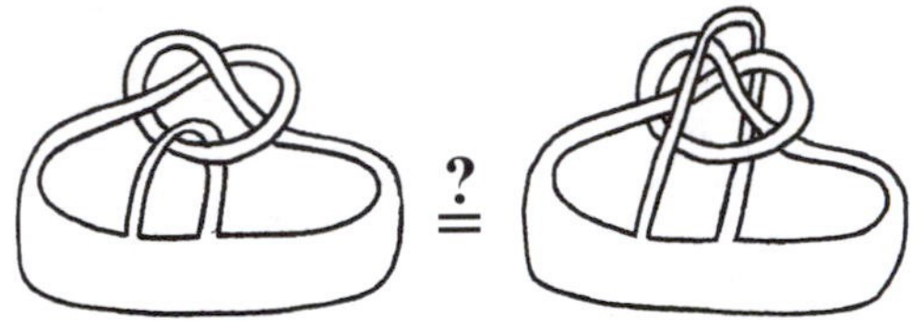

Acknowledgments and Further Reading

These party tricks are familiar to all topologists and appear in many popular books such as [Gard1]. See also [Gard15], chapter 5, and [Tant3]. The waistcoat trick is usually phrased as a challenge to remove the vest without removing the jacket, but you must slip material down the arms of the coat to do this.

19 Tiling and Walking

19.1 Skew Tetrominoes

Here's a trick: Color the cells of the region black and white according to the scheme shown.

Notice that no matter how a skew tetromino is placed on this diagram it will necessarily cover an odd number of black squares. As the diagram is composed of 60 cells, any tiling of it will use 15 skew tetrominoes, and thus will cover an odd number of black squares in total. But the figure possesses an even number of black cells, so no such tiling is possible!

Comment. A similar argument can be used to classify those rectangles tilable solely with tetrominoes of type II.

If an $a \times b$ rectangle is to be tiled with tetrominoes (of any type) we certainly need $4|a \cdot b$. This gives two possibilities:

i) One or both of a and b is divisible by 4.
ii) Neither a nor b is divisible by 4, but both are even.

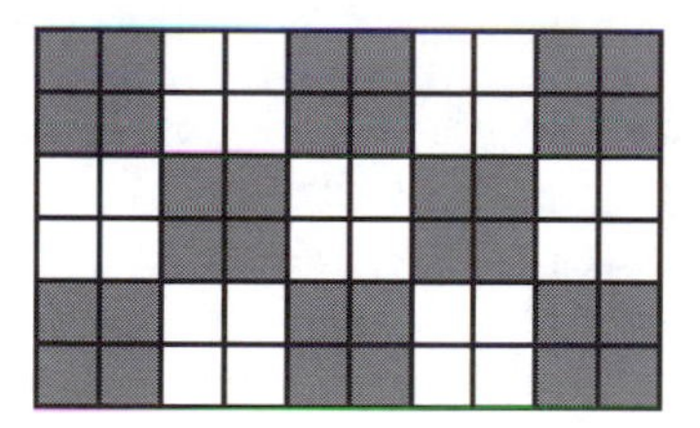

It is easy to see that rectangles satisfying i) are tilable with type II tetrominoes. Consider then, rectangles satisfying ii). Color the cells with the same broad checkerboard pattern as above.

Notice that a tetromino of type II, no matter where it is placed, will cover an equal number of black and white cells. As the total number of black cells differs from the total number of white cells, we conclude then that a tiling with type II tetrominoes is impossible. Only rectangles satisfying i) then are tilable with these tiles.

Challenge. Classify those rectangles that can be tiled solely with tetrominoes of type I, type III or of type IV.

19.2 Map Walking

One could walk an infinite number of closed loops in city A to produce a closed loop shadow journey in city B. A few simple ones are shown here.

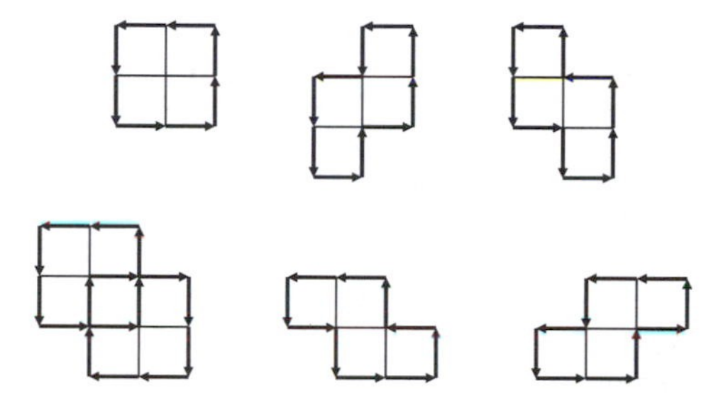

Challenge. These loops work just as well for a person in city B describing motion to a person in city A. Are there any loops that work only "in one direction"?

19.3 Bringing it Together

Look at the skew tetrominoes in the solution to 19.2. If an inhabitant of city A (Albert) were to walk the boundary of one of these tiles, the inhabitant of city B (Beatrice) following the described instructions, will also walk a closed loop path.

The same would occur if Albert were to walk a path that bounds the union of two skew tetrominoes. For example, assume Albert starts at position P on the diagram and walks counterclockwise to position X, then to Y, and finally

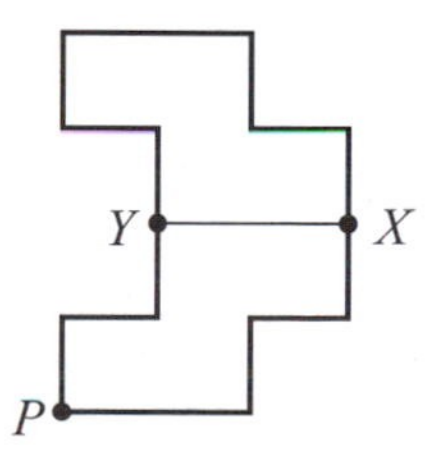

back to P. Beatrice is at the corresponding locations P', X', and Y' in her city at these times.

Now let's reenanct the motions of Albert and Beatrice along this path. First Albert walks clockwise from P to X on to Y, and Beatrice shadows him by walking from P' to X' on to Y'. Suppose Albert deviated from his planned journey momentarily and walked from Y back to X along the short interior path shown. What will Beatrice do? Since Albert just completed the outline of a skew tetromino she will return to X'! Albert, to undo the effect of cutting across the interior path, can follow it back again to Y and continue along the original path planned to return to P. But this completes the loop of a second skew tetromino, and so Beatrice too will complete a second loop. She returns to P'. Dashing back and forth across the interior segment has no effect on the final position of the journeys for either Albert or Beatrice. (All "damage" done is immediately undone.) We conclude, Beatrice, in shadowing Albert's path, will also follow a closed loop.

We can use essentially the same argument to prove that if Albert of city A were to walk the boundary of any (connected) union of (non-overlapping) skew tetrominoes, then Beatrice of city B, following the appropriate instructions, would also follow a closed path. (Use an induction argument on the number of skew tetrominoes.)

Thus to check the potential tilability of the region in question, all we need do is copy the boundary of the region onto a map of city A and see whether the corresponding path in city B is closed. For the region in question, as you may have checked, this is not the case, which means that the region cannot be written as a union of skew tetrominoes. That is, it cannot be tiled! (not even with a mixture of any tiles you came up with in the solution to 19.2!)

Challenge. Is it possible to tile the region below with vertically placed dominoes and the four skew tetrominoes?

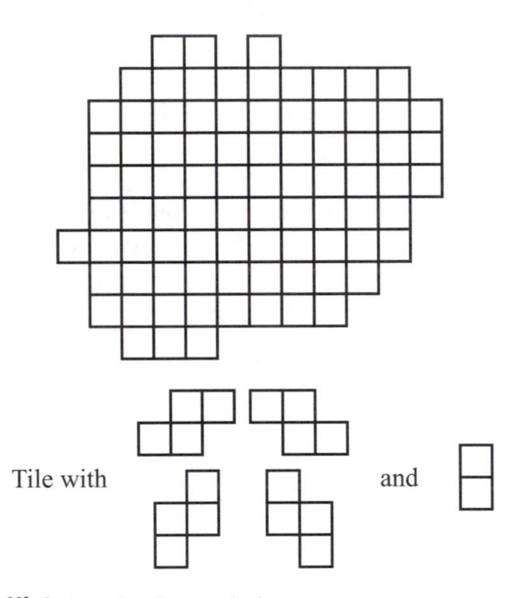

Hint. See the figures below.

Acknowledgments and Further Reading

This chapter is based on James Propp's beautiful and clever article [Prop2] which explores these issues, pushing them even farther, with criteria on regions that are skew-tetromino tilable. (Not all regions that offer closed paths

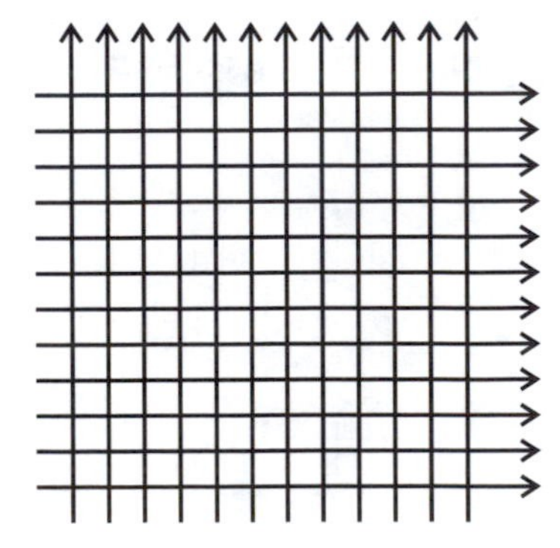

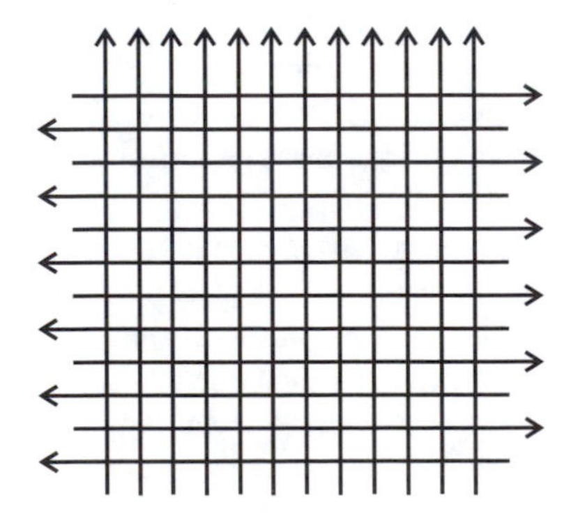

in the two cities are tilable.) His article is based on the work of J. H. Conway and J. C. Lagaris [Conw2]. W. Thurston [Thur] offers new geometric insights and discusses the domino tilability of regions. The classic book on tiling with tetrominoes and other polyominoes is [Golo].

20 Automata Antics

20.1 Basic Ant Walking

Any loop of steps the ant completes must be composed of an even number of steps: Steps taken to the left must be counterbalanced by the same the number to the right, and similarly for the upward and downward steps. See also section 30.1. Since the ant alternates between vertical and horizontal motions, it will always return to that cell from a horizontal direction.

20.2 Ant Antics

Suppose there were a labelling scheme that contained the ant within a confined region. If we let the ant run indefinitely about the grid it will visit some cells infinitely often. (In fact it will return to each of the cells it visits an infinite number of times! Can you see why?) Consider the leftmost top square that the ant visits infinitely often when wandering about the grid. Since the ant alternates between horizontal and vertical motions, section 20.1 dictates that it visits this cell repeatedly from either a vertical or a horizontal direction. In the first case (the second case is similar) these visits can only be from an upward vertical direction (since we chose the **top** leftmost cell). As soon as the ant relabels this cell L, the ant will be forced to leave that square upon its next visit, which contradicts the fact that we chose the **left**most top cell visited. This contradiction shows that no such labelling scheme can exist.

Challenge 1. An ant wanders about a triangular lattice in an analogous way, turning left or right 60° according to the label of the cell it visits. Prove that in this case it is possible to bound the motion of the ant within a finite region.

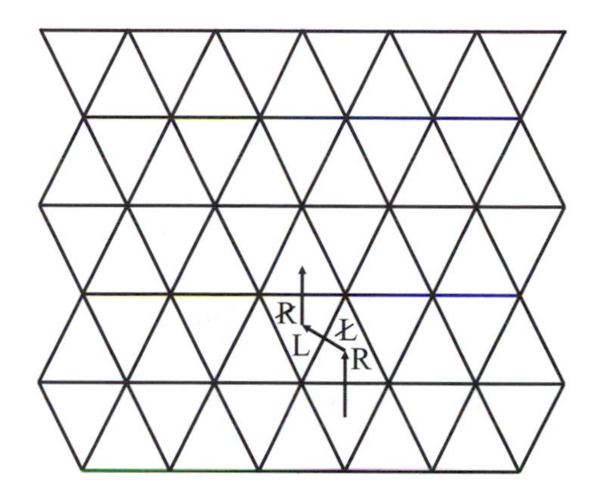

Challenge 2. Prove that it is impossible to bound the motion of an ant wandering about a hexagonal lattice, turning left or right 120° according to the label of the cell it visits.

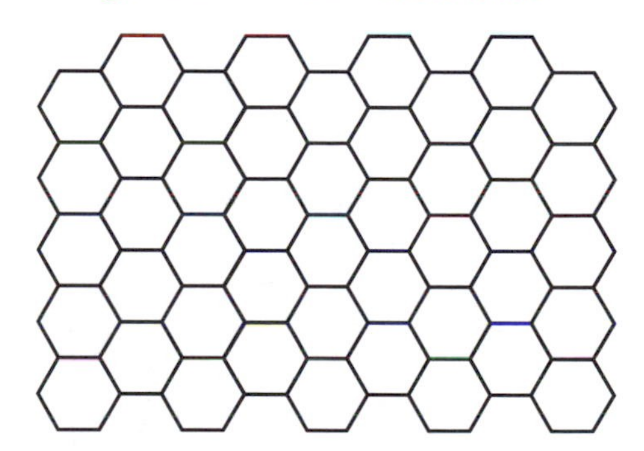

20.3 Ball Throwing

My students at St. Mary's College of Maryland have not yet found a situation where not all people have a turn at throwing the ball (though with 22 people in one group it took quite a while for the ball to make it all the way around!).

After testing the situation on a computer, we feel confident in conjecturing that in all cases a ball thrown within a circle (or an ant wandering about a complete graph) will eventually reach every vertex. As with an ant on a grid, we expect the ant eventually to escape any defined subportion of the graph. A proof of this, however, eludes us. Do you have any thoughts on the matter? Or have you perhaps stumbled on a configuration for which some people fail to have a turn thus proving us wrong?

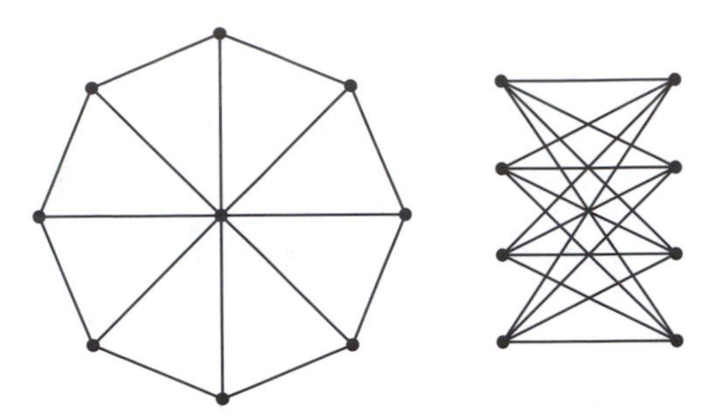

Hard Challenge. Can anything be said (and proven?) about ant motion on other types of finite graphs?

A Note on Ant Motions

The motion of an ant on the infinite plane, divided into a lattice of squares, was first investigated by Chris Langton; see [Lan] or [Stew] for a good introduction. In Section 20.2 we proved the **Cohen-Kong Theorem**: The motion of an ant on a square lattice is unbounded, no matter the initial labelling scheme given to the cells. In other words, it is impossible to contain the ant within a box.

It is very difficult to keep track of the ant's motion by hand, and infeasible to examine its long-term behavior on a sheet of graph paper. We know that the ant will move arbitrarily far from its starting cell given enough time, but we have no indication of the manner in which it will do this. A computer comes to our aid here.

It seems, with all cells initially labelled L, the ant follows three distinct phases of motion. First, for the first few hundred steps, the ant surges back and forth leaving tracks with surprisingly regular symmetry. (This symmetry is most striking with cells colored black and white rather than labelled L and R.) For the next 10,000 steps or so, the ant's motion turns to pure chaos, all symmetry is destroyed, and all rhyme and reason flee the scene—until suddenly, for no discernible reason, the ant settles into systematic periodic motion, drifting off to infinity in a diagonal direction. James Propp, who first noticed this behavior, called it "highway building." This behavior persists even if we change the initial labelling scheme a little by inserting a finite number of R cells. As far as I am aware, no one can yet explain this curious highway-building behavior of the ant.

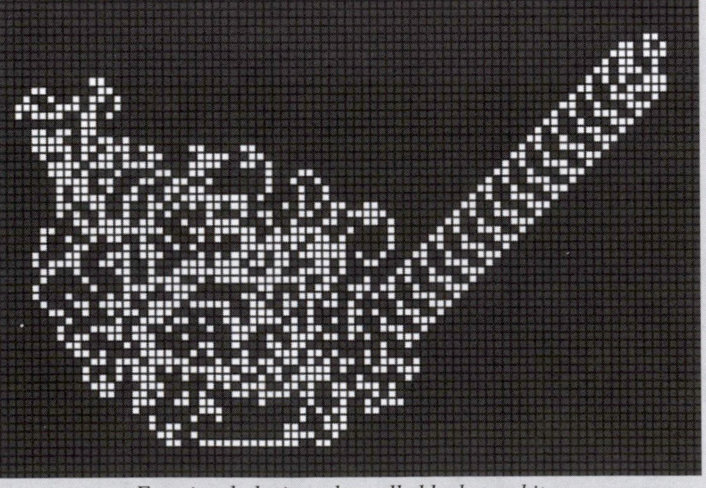

For visual clarity color cells black or white.

Acknowledgments and Further Reading

There is extensive literature on the behavior of ants and "turmites." Check out [Dewd], [Gale2], [Gale3], [Lan], [Prop1], and [Stew]. The behavior of ants on finite graphs was studied by Tom York and Linda Elkins and, briefly, by myself. My thanks to these folks for some key ideas and especially to Tom York for creating the picture of the Langton ant's motion.

21 Bubble Trouble

21.1 Road Building

We need to choose a value of x that minimizes the total road length for such a design.

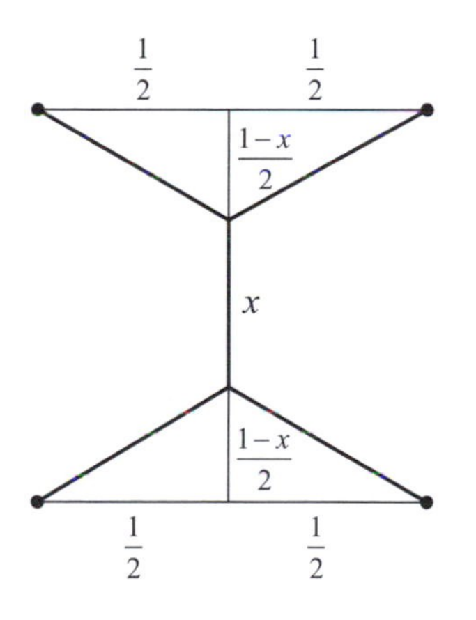

The total length of road is given by:

$$L(x) = x + 4\sqrt{\left(\frac{1}{2}\right)^2 + \left(\frac{1-x}{2}\right)^2}$$
$$= x + 2\sqrt{1+(1-x)^2}.$$

The critical points of this function occur when $L'(x) = 0$, that is,

$$1 - \frac{2(1-x)}{\sqrt{1+(1-x)^2}} = 0$$

or

$$2(1-x) = \sqrt{1+(1-x)^2}.$$

Squaring both sides and solving for x yields

$$x = 1 - \frac{1}{\sqrt{3}}.$$

Using the first derivative test we can check whether this really is a minimum. Thus the minimal length of road possible for such a design is

$$L\left(1-\frac{1}{\sqrt{3}}\right) = 1+\sqrt{3} \approx 2.732.$$

This is certainly better than any of the configurations listed in the problem statement! But is it the most efficient design of all?

The answer turns out to be yes. One can prove it mathematically (see the Note on Steiner Problems) or use a soap solution again. Take two plastic frames and place four upright pegs between them at the positions of the vertices of a square. Dipping this model into soap solution reveals the optimal solution to the problem. (The surface tension in a liquid film always acts to minimize the film's surface area, pulling it to a stable solution.) It is indeed this winged design!

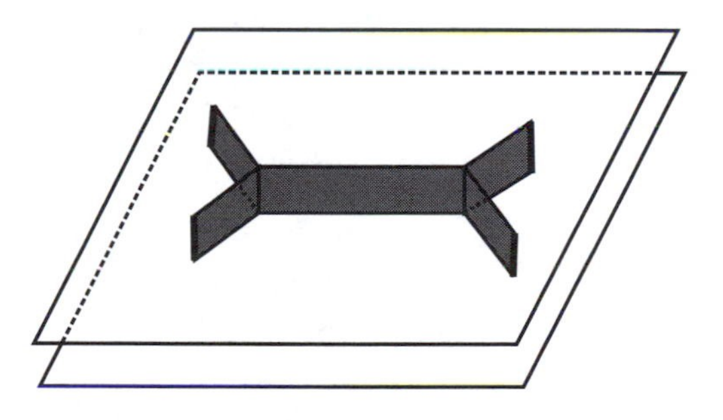

By changing the location and the number of pegs one can experiment with other road-building problems and their solutions.

Challenge. Five towns located at the vertices of a pentagon are to be connected by a road. What is the optimal configuration of the road?

A Note on Steiner Problems

Jacob Steiner, a geometer at the University of Berlin during the early nineteenth century, was the first to pose the problem of connecting three or more given points on a plane via paths of minimal total length. The version with four towns situated on the vertices of a square has become the popular presentation of the problem. Its solution is certainly a surprise.

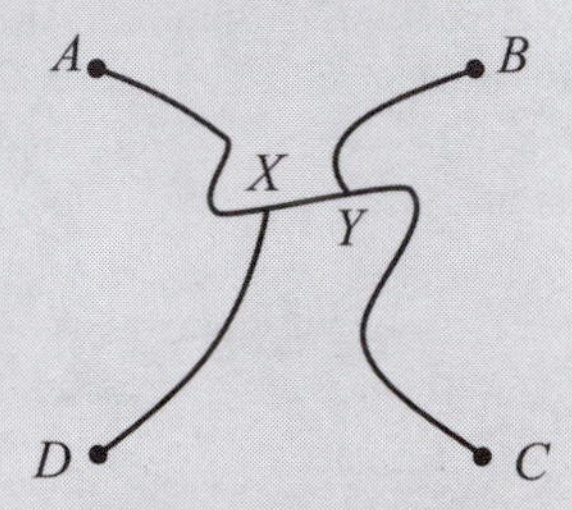

We have made the claim, and verified experimentally, that the winged design offers a solution to the puzzle. (There is another solution, namely a 90° rotation of this one.) Our aim here is to prove this is so mathematically. We first make the assumption that the problem does indeed have a minimal solution in the first place. Our mission is to show that this winged design is it. (*Note*: Proving the existence of a solution turns out to be tricky. We'll just take it as a given here.)

Label the four vertices of the square A, B, C, and D. In any design, there must be a path, call it P, that connects vertex A to vertex C. Having established this, there must also be paths connecting vertices B and D to this path P, intercepting at points X and Y, say. (Note: The points X and Y could be vertices as in the first illustration on page 51, or equal to each other as in the second and third illustrations.)

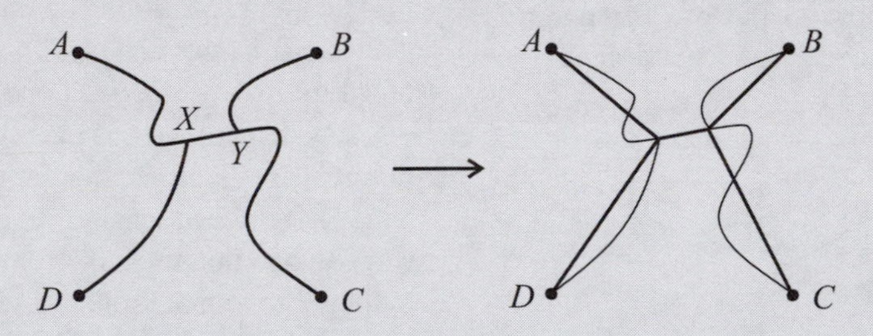

Replacing all intermediate path segments with straight line segments produces a more efficient design, so let's assume we are dealing with a diagram composed of five (or fewer) straight line segments, with two (or one) interior *Steiner points* X and Y. In fact, it is not the case that there is just one Steiner point. If this were so, then the two paths connecting vertices A to C and B to D would have total length at least $\sqrt{2}+\sqrt{2}$ (twice the diagonal length of the square). This is worse than the winged design we presented, which has total path length $1+\sqrt{3}$. So any design with just one Steiner point is not optimal.

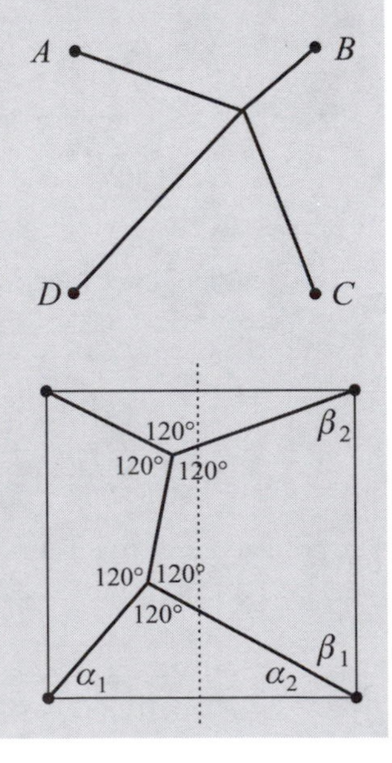

Thus we must be dealing with a diagram composed of two distinct Steiner points and five straight line segments. I now claim that the three lines emanating from each Steiner point are evenly spaced, forming three angles of 120° about each point. If this is so, it then follows that the two Steiner points must be located on opposite sides of the vertical (or horizontal) line of symmetry of the square.

Why is this true? In the diagram on the right $\alpha_1 + \alpha_2 = 60°$. As the point X does not lie on the vertical line of symmetry $\alpha_1 \neq \alpha_2$. In fact, we must have $\alpha_1 > \alpha_2$ (since the point X lies on the circle with line DC as a chord) and so $\alpha_2 < 30°$. Consequently, $\beta_1 > 60°$ and, by analagous reasoning, $\beta_2 > 60°$. But the angles in a quadrilateral sum to 360° and so $120° + 120° + \beta_1 + \beta_2 = 360°$, that is, $\beta_1 + \beta_2 = 120°$. This is a contradiction. The Steiner points thus lie in catercorner quadrants of the square.

The two Steiner points must also be symmetrically placed, for otherwise we could obtain a diagram of smaller total road length by reflecting one half of the picture over to the other side.

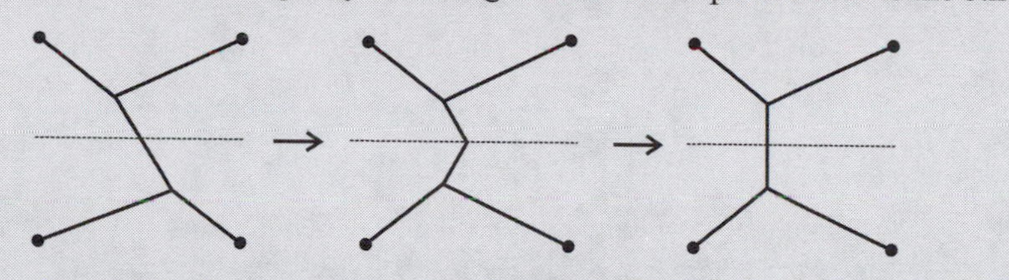

This shows that in fact the two Steiner points, in our picture, lie on the vertical line of symmetry of the square. This reduces the analysis of the problem to that of the diagram in the hint, for which we know an interior segment of length of $1 - 1/\sqrt{3}$ units yields the optimal solution.

To prove our claim we need to use some other results regarding minimal total lengths.

Result 1. *Let P and Q be two points on one side of a line L. If R is a point on L such that PR + RQ is a minimum, then PR and RQ make equal angles with L.*

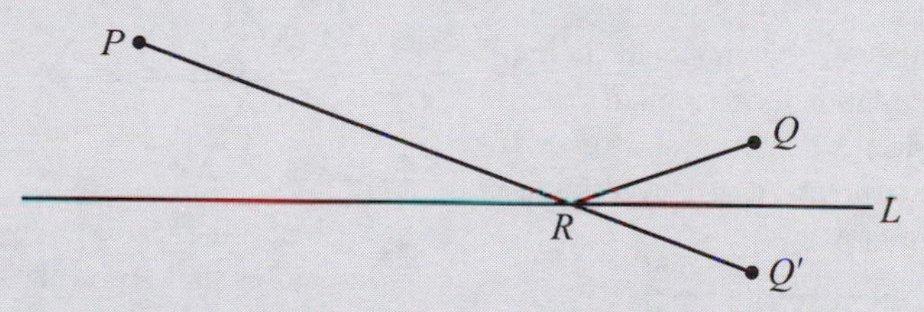

This follows by looking at the image point Q' of Q under a reflection about the line L. Clearly the shortest path connecting P to Q' (a straight line) yields a path from P to Q via a point R on L with minimal sum $PR + RQ$. This path offers the equal angles described.

Result 2. *Let P and Q be two points on one side of a tangent line to a circle C (with the circle on the other side). If R is a point on C such that PR + RQ is a minimum, then PR and RQ make equal angles with the tangent line L to the circle at R.*

Draw an ellipse with foci P and Q that passes through the point R. It is not difficult to see that the line L must also be tangent to this ellipse and that R is in fact a point on L that minimizes the sum $PR + RQ$. By result 1, the claim follows.

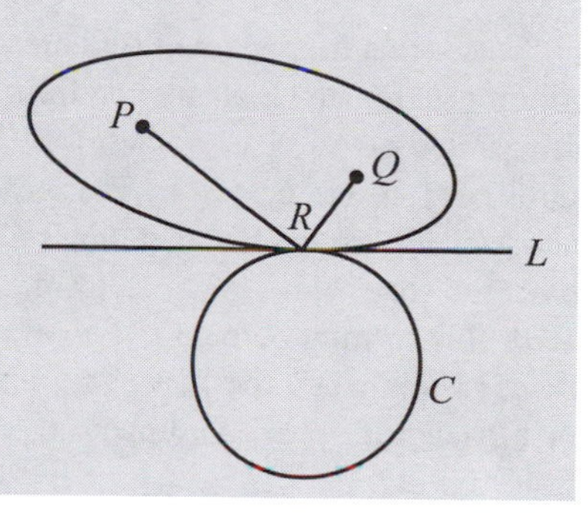

Assume that we have a minimal solution to the four-vertex problem with two Steiner points. Without loss of generality we are working with a diagram of the form shown on the next page.

Suppose the length of the segment connecting D to X is r. Draw a circle of radius r with center D. Because we

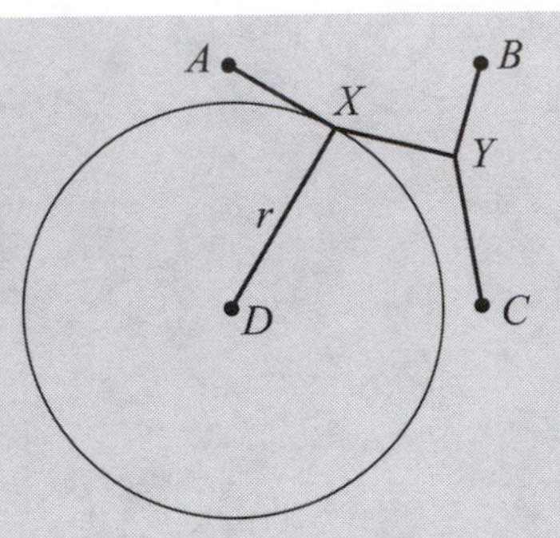

are working with a minimal total length solution, it must be the case that X is a point on this circle that minimizes the length $AX + XY$. By result 2 it follows that $\angle DXA = \angle DXY$. Similarly, by drawing a circle about A and one about Y, we have $\angle AXD = \angle AXY$ and $\angle YXA = \angle YXD$. Thus all three angles are equal, proving the claim for the Steiner point X. A similar proof works for Y.

This completes the proof that the winged diagram offers an optimal solution to the four vertex Steiner problem.

21.2 Higher Dimensional "Road Building"

In an analogy to the two-dimensional Steiner problem, the soap solution forms a small square of film hovering in the center of the cube. If you gently tap the structure, this central square can flip to a different plane!

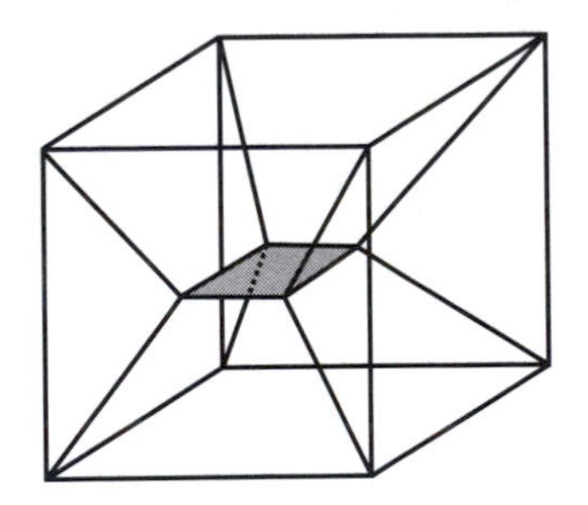

Notice that four edges of film meet at every interior vertex and that any two films meeting at an edge do so at an angle of 120° (sound familiar?). It was proven by F. J. Almgren, Jr. and J. E. Taylor in 1976 that this is always the case for soap film structures [Almg], [Hild]. With this in mind, would you care to predict what results when the frame of a tetrahedron or a triangular prism is dipped in soap solution? Try it!

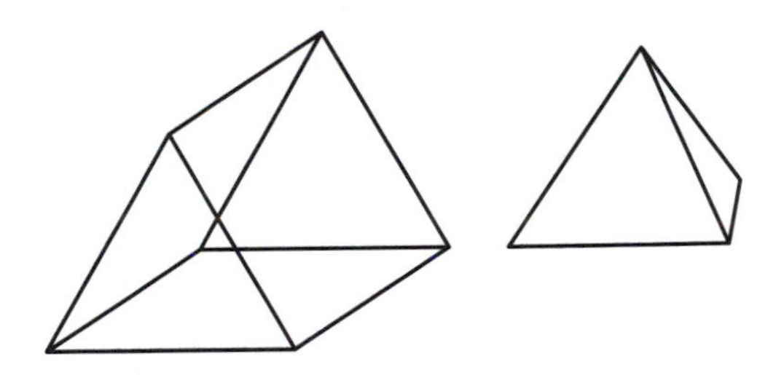

It is great fun to experiment with a variety of wire structures and see what minimal surfaces result when they are dipped in soap solution. Try for example the boundary of a Möbius band (see sections 8.1 and 10.2). This is formed from a loop of wire twisted to form a squashed figure eight.

There are at least two minimal surfaces associated with this figure. Can you create them both?

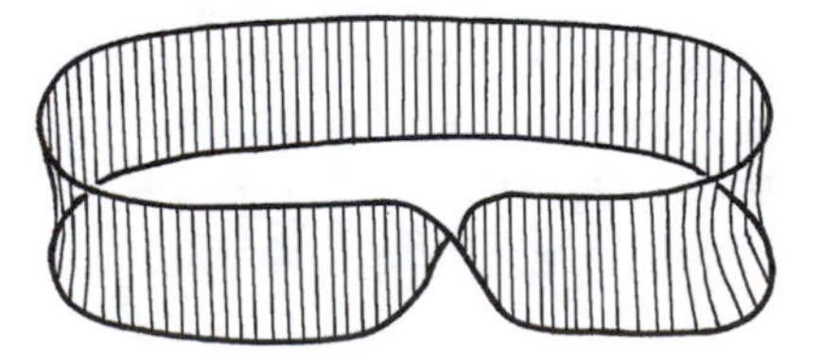

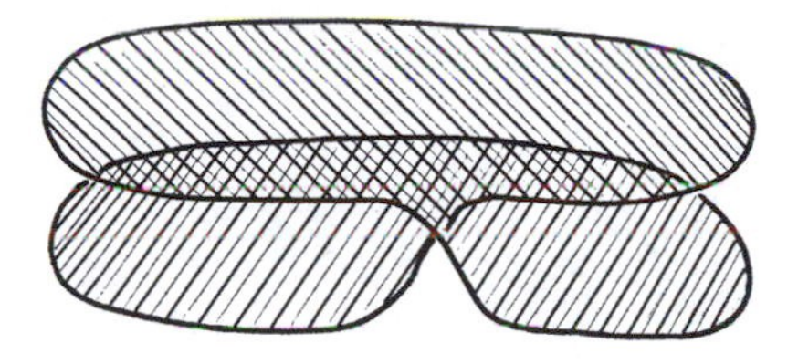

Here are some more exotic wire frames to dip into soap solution. How many different surfaces can you create with them?

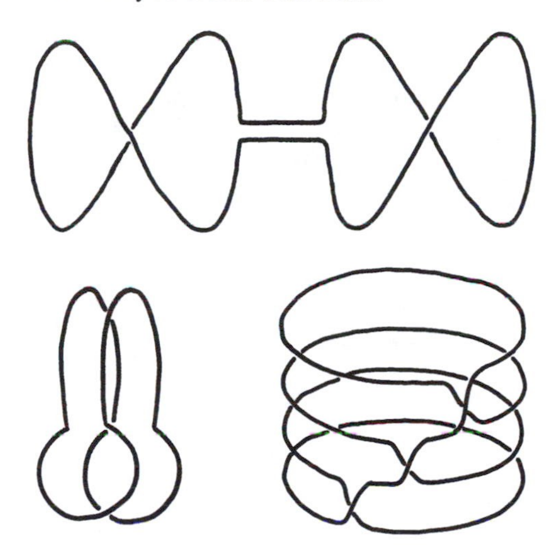

21.3 Donut Bubbles

This too is impossible, but it wasn't proven so until very recently. In 1995 J. Hass et al. [Hass] proved that if two bubbles enclose the same volume of air, then the familiar double-bubble configuration is the only stable double-bubble configuration. (See [Hass1] and [Pete4].)

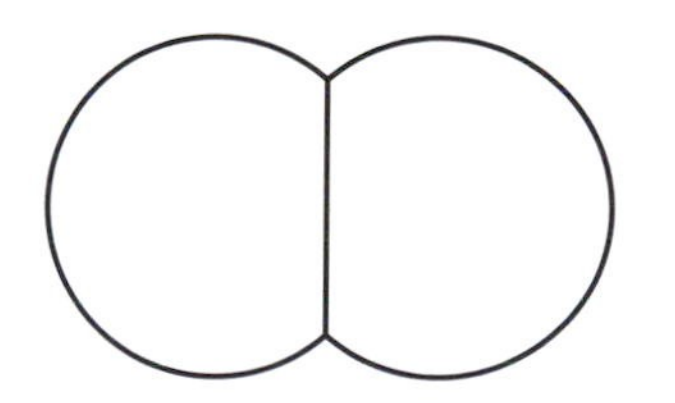

Recently M. Hutchings, F. Morgan, M. Ritoré and A. Ros announced their proof of the general double- bubble conjecture: the standard double-bubble configuration again is the only stable bubble configuration enclosing two different sized pockets of air. (See [Hass2].)

Challenge. Can one make toroidal bubbles as part of a *triple*-bubble configuration?

Acknowledgments and Further Reading

Steiner analysis and interesting soap film constructions can be found in Richard Courant and Herbert Robbins' classic book [Cour]. Also look at Stefan Hildebrandt and Anthony Tromba's stunningly beautiful book [Hild], and Ivars Peterson's books [Pete2], [Pete4]. Other intriguing soap film questions and experiments can be found in Frank Morgan, Edward Melnick and Ramona Nicholson's fabulous article [Morg].

22 Halves and Doubles

22.1 Freaky Wheels I

This freak is an artifice of compound motions. In the first instance, with two wheels rotating simultaneously, after just half a turn the two arrows are aligned pointing downward. If at this instant we were to rotate the two wheels 180°, as a pair the two arrows would be aligned pointing upward. This is in effect what occurred within the second scheme of motion. One wheel moves as it rotates and in the process turns itself upside down.

Challenge. What happens if the wheels are of different sizes?

22.2 Freaky Wheels II

The problem lies in the fact that the little wheel slides along the track as it rotates. Galileo analyzed this problem as follows: Imagine instead two square "wheels." The big wheel pivots about its corner to perform one quarter of a rotation, lifting the little wheel above its track as it rotates.

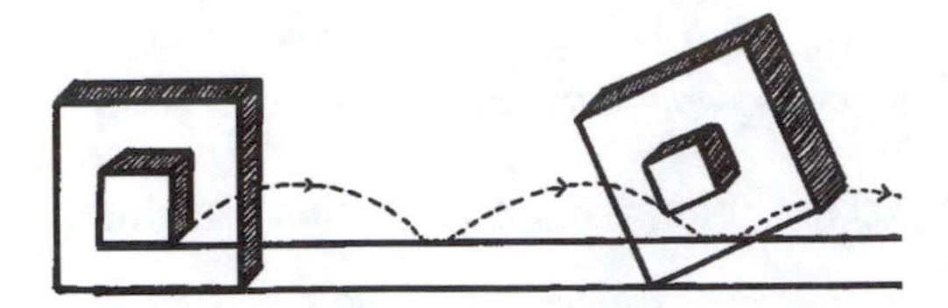

For polygonal wheels with a greater number of sides, the little wheel performs numerous smaller jumps as it rotates. In the limit of a circular wheel, this corresponds to the little wheel sliding forward as it turns.

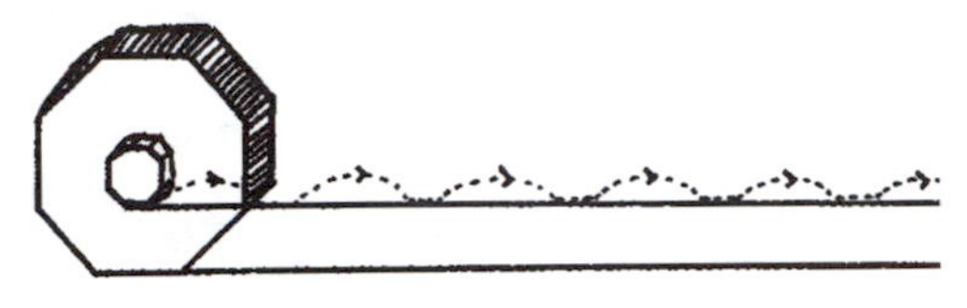

22.3 Breaking a Necklace

Draw an arrow through the center of the circle, passing in-between pearls. Let x be the number of black pearls to the right of this arrow

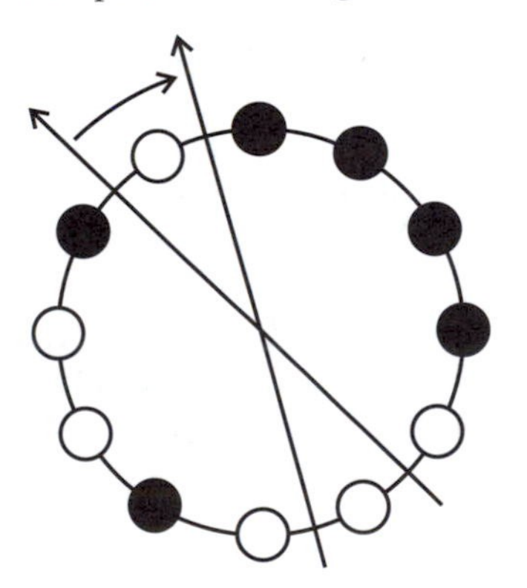

(thus there are $6 - x$ black pearls to its left.) If we rotate the arrow one position over, the value of this number x could change by ± 1 or not at all, depending on whether we pick up or lose extra black pearls in the process. Moving six positions over brings the arrow parallel to its initial position but pointing in the opposite direction. Thus the number of black pearls to its right is $6 - x$.

Thus we can move from having x black pearls to the right to having $6 - x$ black pearls to the right, changing the numbers by ± 1 at a time (for example, from four pearls down to two for the picture above). There must be an intermediate position with three black pearls on each side of the line.

Challenge. A necklace contains $2n$ black pearls and $2m$ white pearls. Can you prove that one half of the necklace contains precisely n black and m white pearls?

Comment. This problem generalizes to necklaces containing more than two types of pearls. Suppose a necklace contains r types of pearls, with $2n$ of each type. Then with just r cuts if r is even ($r + 1$ cuts if r is odd) it is possible to share the pearls equally between two pirates. Try playing with it a few times to see that this is true!

In the note below I will prove this claim for the case of $r = 3$ types of pearls (the proof generalizes to higher values of r). Unfortunately it is a very complicated proof. Perhaps you can develop an easier one.

A Note on the Borsuk-Ulam Theorem

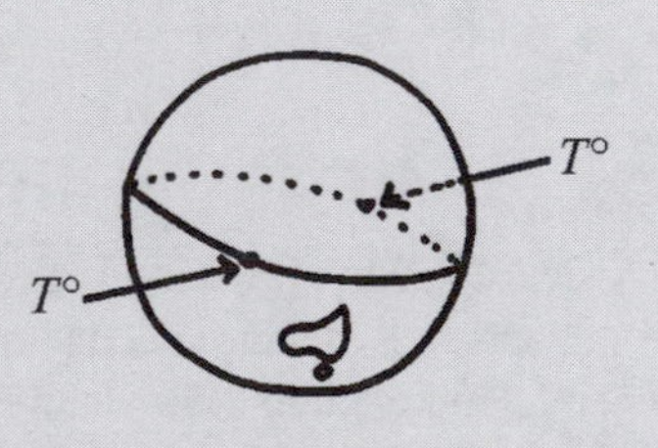

(*Warning:* Lots of calculus!) Let's start with a bold claim: At any instant there are two points on the earth's equator, directly opposite one another, with precisely the same air temperature.

This is a common interpretation of a more general (rigorously stated!) result: For any continuous function on a circle, f: Circle $\to \mathbb{R}$, there exists an angle θ such that $f(\theta) = f(\theta + 180°)$. This result follows from a clever use of the *Intermediate Value Theorem,* which states that a continuous function assumes all values between any two given function values.

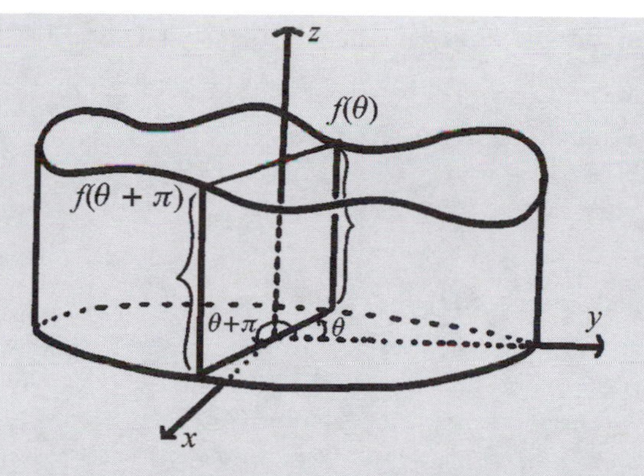

Thus, if you know that a continuous function f adopts a value 4 at one point and 5 at another, there must be a point x where $f(x) = 4.2$ say.

For a continuous function f on a circle let $g(\theta) = f(\theta) - f(\theta + 180°)$. Then $g(0°) = -g(180°)$ and so there must be an angle θ between 0° and 180° with $g(\theta) = 0$.

Challenge 1. Let θ be any angle a rational multiple of 360°. Prove that at any instant there are two points on the earth's equator that angle apart with precisely the same air temperature. (**Hint.** Practice with $\theta = 120°$ first.)

The surprising thing is that this result generalizes to higher dimensions. For example, at any instant, there are two points, directly opposite each other on the Earth (*antipodal points*) with precisely the same temperature *and* barometric pressure! This is the Borsuk-Ulam Theorem ($n=2$ case): *For any continuous map* f : Sphere $\rightarrow \mathbb{R}^2$ *there must exist two antipodal points with the same function value.*

This theorem is valid for higher dimensional spheres as well (see section 10.2). For example, any continuous map from a sphere in four-dimensional space into $\mathbb{R}^3$ must have a pair of antipodal points with the same function value ($n = 3$ case), and so on. For a proof of the Borsuk-Ulam theorem in all dimensions, see the advanced book [Mass].

We will use the $n = 3$ version of the theorem to solve the necklace cutting problem of section 22.3 in the $r = 3$ case. See section 22.4 for another application of the Borsuk-Ulam Theorem.

(*Warning:* This really is tricky!) Suppose our necklace contains $2n$ black, $2n$ white, and $2n$ grey pearls. Let f_1 be a function whose graph consists of horizontal line segments of height 1 along every position occupied by a black pearl as we sweep around the necklace from angle 0° to angle 360°. Similarly construct functions for f_2 and f_3 for the white and grey pearls respectively.

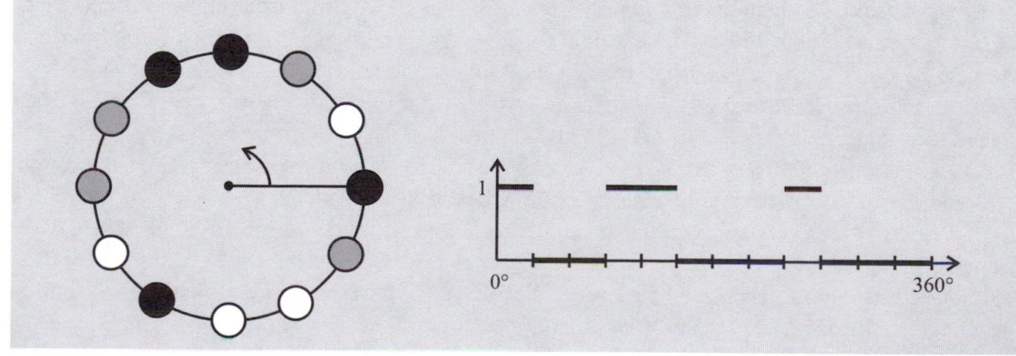

A sphere sitting in four-dimensional space is given by (section 10.2):

$$S^3 = \{(x, y, z, w) | x^2 + y^2 + z^2 + w^2 = 1\}.$$

For a point (x, y, z, w) on such a hyper-sphere let

$$X(x,y,z,w) = \operatorname{sign}(x)\int_0^{x^2} f_1(t)\,dt + \operatorname{sign}(y)\int_{x^2}^{x^2+y^2} f_1(t)\,dt$$
$$+\operatorname{sign}(z)\int_{x^2+y^2}^{x^2+y^2+z^2} f_1(t)\,dt + \operatorname{sign}(w)\int_{x^2+y^2+z^2}^{1} f_1(t)\,dt,$$

where

$$\operatorname{sign}(x) = \begin{cases} +1 & \text{if } x > 0, \\ 0 & \text{if } x = 0, \\ -1 & \text{if } x < 0. \end{cases}$$

Define $Y(x, y, z, w)$ and $Z(x, y, z, w)$ in the analogous way with f_1 replaced respectively by f_2 and f_3. Thus we have a continuous function $f : S^3 \to \mathbb{R}^3$ given by

$$f(x, y, z, w) = (X(x, y, z, w), Y(x, y, z, w), Z(x, y, z, w)).$$

Notice that with the presence of the sign functions, f is an odd function, that is, $f(-x, -y, -z, -w) = -f(x, y, z, w)$ for all points (x, y, z, w). By the Borsuk-Ulam theorem there must exist a particular pair of antipodal points (x, y, z, w) and $(-x, -y, -z, -w)$ for which the function values are the same. Because we are dealing with an odd function this function value must be zero. Thus we have found a point (two actually) where $f(x, y, z, w) = 0$. For this point, therefore,

$$0 = \operatorname{sign}(x)\int_0^{x^2} f_1(t)\,dt + \operatorname{sign}(y)\int_{x^2}^{x^2+y^2} f_1(t)\,dt$$
$$+\operatorname{sign}(z)\int_{x^2+y^2}^{x^2+y^2+z^2} f_1(t)\,dt + \operatorname{sign}(w)\int_{x^2+y^2+z^2}^{1} f_1(t)\,dt,$$

$$0 = \operatorname{sign}(x)\int_0^{x^2} f_2(t)\,dt + \operatorname{sign}(y)\int_{x^2}^{x^2+y^2} f_2(t)\,dt$$
$$+\operatorname{sign}(z)\int_{x^2+y^2}^{x^2+y^2+z^2} f_2(t)\,dt + \operatorname{sign}(w)\int_{x^2+y^2+z^2}^{1} f_2(t)\,dt,$$

$$0 = \operatorname{sign}(x)\int_0^{x^2} f_3(t)\,dt + \operatorname{sign}(y)\int_{x^2}^{x^2+y^2} f_3(t)\,dt$$
$$+\operatorname{sign}(z)\int_{x^2+y^2}^{x^2+y^2+z^2} f_3(t)\,dt + \operatorname{sign}(w)\int_{x^2+y^2+z^2}^{1} f_3(t)\,dt.$$

Cut the necklace at angles 0, x^2, $x^2 + y^2$, and $x^2 + y^2 + z^2$, cutting through pearls if necessary (just four cuts). Some of the numbers sign(x), sign(y), sign(z), and sign(w) must be positive and others negative (the integrals sum to zero.) Let one pirate take those sections of string corresponding to integrals with positive coefficients, the other with the negative coefficients. They each thus end up with the same amount of each pearl.

Do we need to cut through the pearls? Actually, no. Suppose some black pearls are cut through. As each person receives a whole number of black pearls, not just one pearl is cut. If two black pearls are cut then one can steer the cuts to the side and avoid cutting them. If three pearls are cut through, we can steer one cut aside, adjusting the other two appropriately, and reduce the situation to two cut pearls, and then none. Similarly the case of four cut black pearls can be reduced to that of three, then two and then none.

Thus we have proven that at most four cuts are needed to divide a necklace with three types of pearls between two pirates!

Challenge 2. Prove the more general necklace subdivision problem with $r > 3$ types of pearls.

22.4 Congruent Halves

Challenge. The "L" shape can be subdivided into 2, 3, 4, 6, 8, and 27 congruent pieces. Show that it can also be subdivided into 14, 17, and 25 congruent pieces! See [Gard19].

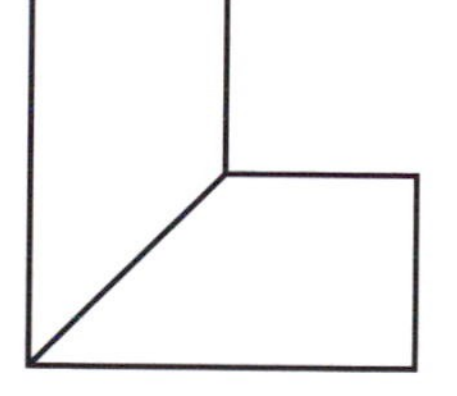

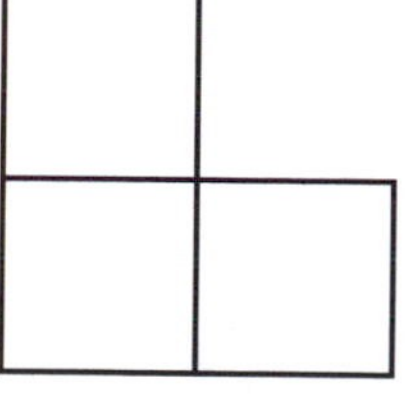

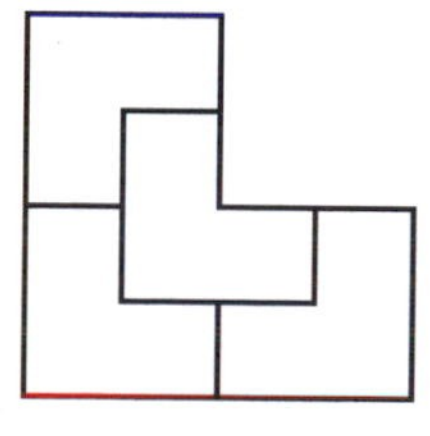

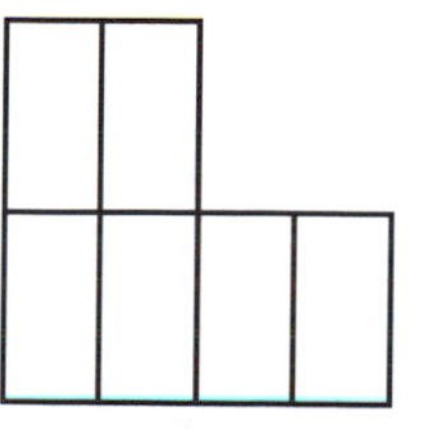

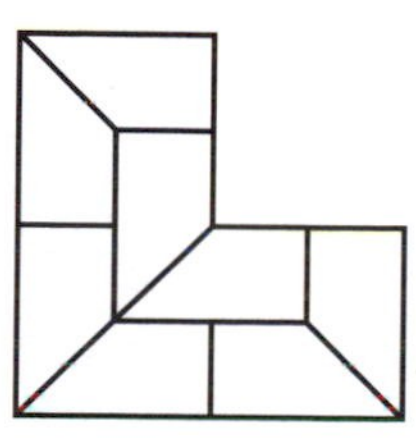

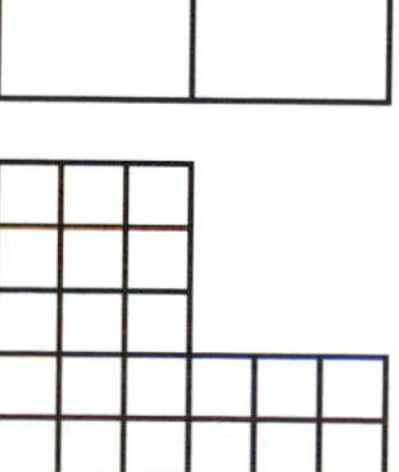

A Note on Cutting Shapes in Half

The Intermediate Value Theorem shows that any (finite) planar region can be subdivided into two equal halves by a straight line positioned at any fixed angle. Simply imagine sliding the line across the figure, from a position with all the area of the region sitting to the right of the line, to one with all the area sitting to the left. There must be some intermediate position where the area is split precisely in two. (Notice that the line at this angle is unique.)

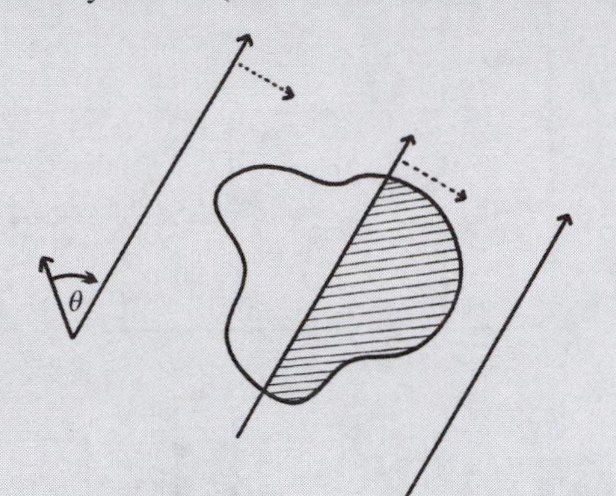

Suppose now we are given two regions on the plane. Does there exist a single line that simultaneously divides the area of each region into two equal parts? Surprisingly the answer is yes. This famous result is known as the **Two Pancakes Theorem**.

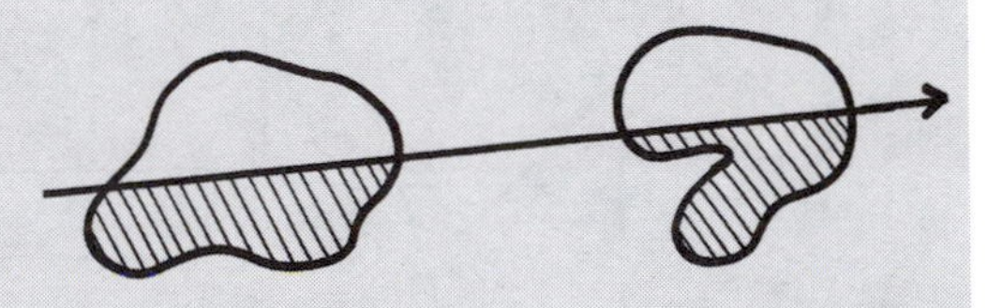

For each angle θ there is a unique line (let's make it an arrow) at that angle that subdivides the first region into two equal parts. This line may miss the second region completely, or cut it into two unequal areas. Let's measure how successful that line is at splitting the second region by defining a function $f(\theta)$ to be the amount of area of the second region to the right of the line minus the amount of area to its left. Ideally we want to find a line at an angle θ such that $f(\theta) = 0$.

The Intermediate Value Theorem guarantees that we'll find such a line! Notice that $f(0°)$, whatever value it has, equals $-f(180°)$;the angles represent the same line but pointing in opposite directions. Because f varies continuously as the angle varies, there must be a position where $f(\theta) = 0$.

The curious thing is that the second region doesn't even need to be an area. It could, for example, be a piece of string whose length we would like to slice in half (but not necessarily into two pieces!).

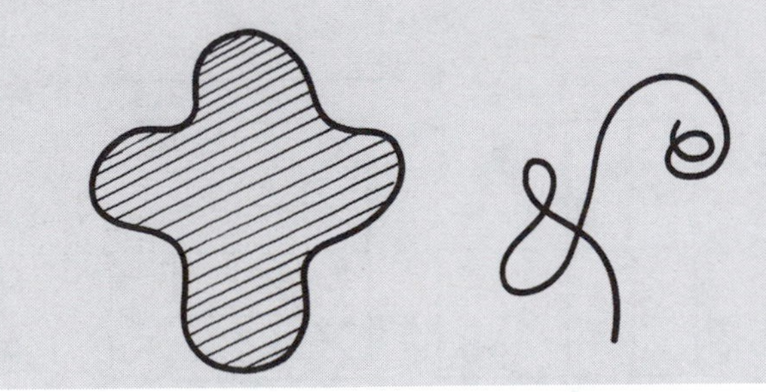

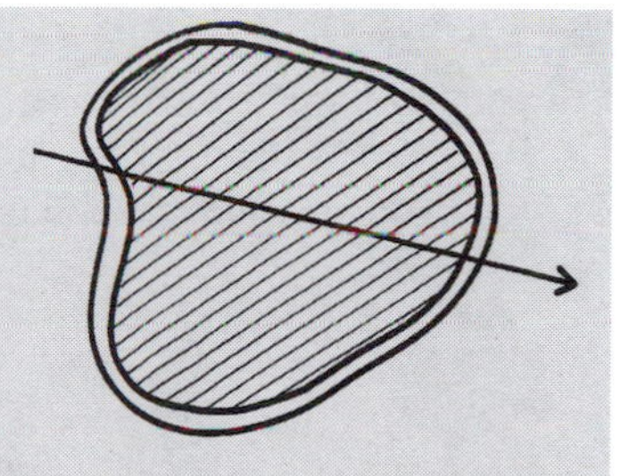

This piece of string could even be the perimeter of the first region under consideration. Thus we have proven what I call the *Pizza Slicing Theorem. It is possible, in a single straight line cut, to slice a pizza, no matter its shape, into two equal portions with each portion containing the same amount of crust.*

Moving up a dimension, imagine a volume sitting in 3-space. Place a small sphere at some position to the side. Now, any point on this sphere determines a ray passing through the center of the sphere and pointing up to the direction of the point. Call this ray L. A given plane orthogonal to L might or not divide that volume into two equal parts. However, sliding the plane up and down the ray L must eventually yield a plane, still orthogonal to L, that does. The Intermediate Value Theorem dictates so.

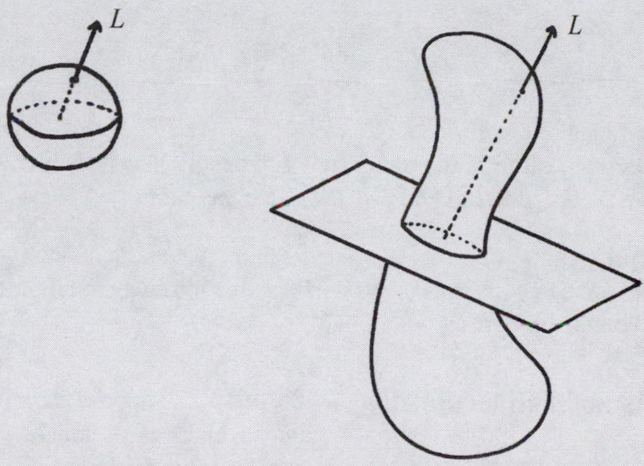

We will now attempt to generalize the Two Pancakes Theorem in 2-space to a three-volume theorem in 3-space. Suppose three volumes are sitting in 3-space: the Eifel Tower, the planet Neptune, and a pencil. I claim there is a single plane that will simultaneously slice each of these three volumes in half! As before, a small reference sphere is placed to one side of these volumes. Let P be a point on this sphere and L the ray that passes through the center of the sphere and up through P. As we have just seen, there is a plane orthogonal to this ray that slices at least the first volume into two halves of equal volume. The other two volumes might or might not be cut.

Let $f_2(P)$ be the volume of region II on one side of the plane (the side containing the direction of the arrow) minus the volume of region II on the other side of this plane. Define $f_3(P)$ similarly. Thus, every point P on a sphere determines two function values $f_2(P)$ and $f_3(P)$. So we really have a function on a sphere, $F: S^2 \to \mathbb{R}^2$, given by

$$F(P) = (f_2(P), f_3(P)).$$

Notice that

$$f_2(-P) = -f_2(P), \quad f_3(P) = -f_3(P)$$

because the antipodal point $-P$ determines the same plane as the point P but the volumes of regions II and III on either side of this plane have switched places.

The Borsuk-Ulam theorem ($n = 3$ case) says there must be a particular point P on the sphere with

$$f_2(P) = f_2(-P), \quad f_3(P) = f_3(-P)$$

and hence with

$$f_2(P) = 0, \quad f_3(P) = 0.$$

Thus the plane associated to this point P divides not only the first volume into two pieces of equal volume, but also regions II and III as well!

This result is known as the **Ham Sandwich Theorem:** It is possible, in a single planar slice, to divide two pieces of bread and a slab of ham each into two pieces of equal volume.

Challenge. A single irregular blob sits in 3-space. Is there a single plane that simultaneously divides its volume into two equal pieces, its surface area into two equal pieces, and one of its "equators" into two equal lengths?

Acknowledgments and Further Reading

The Freaky Wheel problems, classic paradoxes, are discussed in Bryan Bunch's wonderful book [Bunc] and in [Tant2]. The pirate necklace problem appears in [Toti]. Planar dissection problems abound in recreational mathematics books and journals; see for example [Gard9], [Gard18].

23 Playing with Playing Cards

23.1 A Pastiche of Card Surprises

Surprise 1. Move all the red cards from the first pile to the second, and all the black cards from the second to the first. Each pile will still contain 26 cards, thus the number of cards transferred must be equal!

Surprise 2. Suppose the pile of 32 cards contains x black cards and $32 - x$ red cards and the other pile contains y black cards and $20 - y$ red cards. Then

$$x - (20 - y) = x + y - 20$$
$$= 26 - 20 = 6$$

no matter the values of x and y! *Note:* Surprise 1 can be viewed as a special case of this trick.)

Surprise 3. The same reasoning as for Surprise 1 works here. There is nothing special about the number ten, and this trick works even if the two initial piles are unequal in size!

Surprise 4. There's not much to this surprise. Experiment to see why it works.

Surprise 5. Initially all the aces, for example, are spaced 13 cards apart. Cutting the deck once, 13, or even 1300 times is not going to alter the cyclic distribution of these cards: They will always be so spaced. (Imagine a wrap-around effect along the deck, where the bottom card is regarded as adjacent to the top card.) The aces will thus be grouped together in a single pile of four when dealt.

Surprise 6. Let M be the magic sum. A total of $40 - M$ cards is dealt to "top up" each selected card to a value of 10. Thus on the bottom of the pile are placed

$$8 + 4 + (40 - M) = 52 - M$$

cards. This means there are M cards on top, the Mth one being the original bottom card.

Surprise 7. Notice where the cards of the middle column are placed when the pile is redistributed as seven rows of three. This maneuver

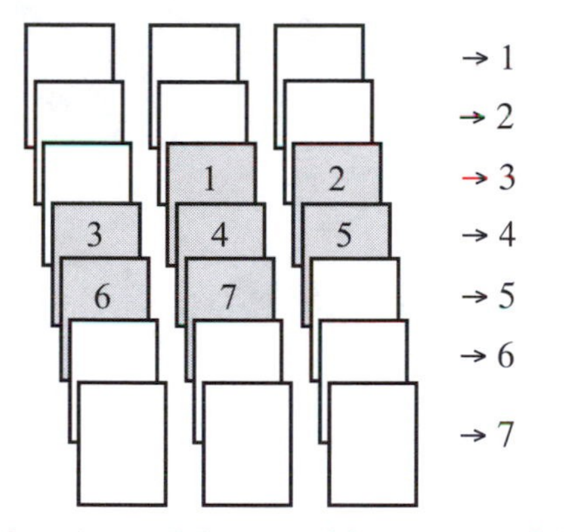

transfers the card from position 1 or 2 of the selected column into the third position of a column; the card from position 3, 4, or 5 of the selected column into the fourth position of a column; and the card from position 6 or 7 into the fifth position:

$$\left.\begin{matrix}1\\2\end{matrix}\right\}\to 3$$

$$\left.\begin{matrix}3\\4\\5\end{matrix}\right\}\to 4$$

$$\left.\begin{matrix}6\\7\end{matrix}\right\}\to 5$$

Repeating this transformation once more ensures that the selected card will be in the middle position of a column, position 4. As soon as your friend indicates to you the appropriate column, you know his or her chosen card.

Challenge. This trick also works with just 15 cards in five rows of three, and 30 cards in ten rows of three. In both cases you can identify the chosen card within two transformations as above. In a game of 33 cards (11 rows of three) you need one extra maneuver in order to pin it down. For which n will a version of this trick work with $3n$ cards, n rows of three? How long will it take to pin down the chosen card in each case?

You can also play this game with four columns of cards, always picking up the selected column second. With how many cards should you play?

23.2 Curious Piles

Hall's Marriage Theorem states that if you are given n sets $A_1, A_2, \ldots, A_n$ with the property that the union of any k of them ($1 \leq k \leq n$) contains at least k distinct elements, then it is possible to select n distinct objects, one from each set. (*Note:* The condition placed on these sets is not trivial. Potentially some sets could be empty or some elements could appear in more than one set. The next "Note" explores this theorem and offers a proof.)

Hall's Marriage Theorem guarantees the means to select four cards of distinct suits, one from each of four distinct piles of 13 cards, or 13 cards of distinct denominations, one from each of 13 piles of four cards. In the latter case, think of each pile as a set containing one, two, three, or four elements: the distinct denominations that appear in that pile. Notice that among any k piles of cards at least k distinct denominations appear (there are at least that many distinct denominations among any $4k$ cards). Thus the existence of one card of each denomination spread out across 13 piles is assured by the theorem.

Pulling in this big theorem is comforting—but not completely satisfactory. Although it assures us that these feats can always be accomplished, it gives absolutely no indication of how to perform them! Trial and error works well enough for the simple four-pile problem, but not with 13 piles. Is there a general procedure for pulling out a selection of distinct denominations from 13 piles of four cards? Fortunately there is.

Choose an arbitrary pile containing an Ace, then one containing a two, and so on, doing this for as long as possible until you get stuck. Place these selected piles in a row. Suppose you have selected a total of ten piles, containing an Ace, and two through ten, respectively. This scenario leaves three untouched piles. If any of the 12 cards in those three piles is a Jack, Queen, or King" then you are not really stuck; you can continue a little further (but perhaps not in the usual sequential order). If you are truly stuck, select any card in, say, the 11th pile, and go to one of the ten piles that corresponds to the number of that card. Thus if you select a three you go to the third pile. Turn a card in that third pile 90° (to make it conspicuous) and move to the pile that corresponds to the number of that turned card. Keep doing this to create a chain of piles and turned cards, until you eventually hit upon a pile that contains a card not in the initial list; in our case, until we come upon a Jack, Queen, or King. (*Note:* One can prove that you wil! not fall into a closed loop of choices.) Now shift all the piles one place back along the chain of piles to obtain a configuration that allows you to add one more pile to the initial list of ten. Repeat this process until you solve the puzzle completely!

Challenge. A deck of cards is shuffled and dealt into 26 piles of two cards. Is it possible to select a black Ace from one pile, a Red ace from another, a black two from a third, a red two from a fourth, and so on all the way down to a black King and a red King from the two remaining piles?

A Note on Hall's Marriage Theorem

A statement of Hall's Marriage Theorem was presented in section 23.2. The reference to marriage arises from the following (terribly dated) interpretation of the theorem: Suppose n *women each list the names of men they would like to marry. As long as any r women mention at least r distinct names among them,* $1 \leq r \leq n$, *then it is possible to find satisfactory marriage arrangements for all.* (Assuming the men are willing!) Ian Anderson presents a "constructive" proof in his book [Ande]. It makes the constructive method described in Section 23.2 rigorous. The simplest proof I know of this result is by induction on the number of women n.

If there is just a single woman with at least one name on her list, then she can satisfactorily marry. Thus the statement is true for $n = 1$. Suppose the statement is true if $n = 1, 2, 3, \ldots,(k-2)$ or $(k - 1)$. Consider the case of $n = k$ women with lists satisfying the property described above. We will consider two cases.

Case I. Suppose these women's lists satisfy the stronger condition that among any r lists, $r < k$, at least $(r + 1)$ distinct names are mentioned. Select one woman. She has at least two names on her list ($r = 1$ case). Have her marry one of these men (Poindexter). This leaves $k - 1$ women to marry.

The lists possessed by these $k - 1$ women have the property that among any r of them ($1 \leq r \leq k - 1$) at least $r + 1$ distinct names are mentioned. One of these names could be Poindexter's who is no longer available for marriage. Thus the best we can say is that among any r lists, at least r distinct names of available men are mentioned. This is all we need to invoke the induction hypothesis for these remaining $k - 1$ women. Thus we have a means to marry all k women in this scenario.

Case II. Suppose this stronger condition does not hold. Thus there is a subgroup of r_0 women ($1 \leq r_0 < k$) such that among their lists precisely r_0 distinct names are mentioned. By the induction hypothesis (for $n = r_0 < k$ case), we can marry these women. This leaves $m = k - r_0$ women to consider.

Is it true that among these m women's lists any r of them ($1 \le r \le m$) mention at least r distinct names of available men? The answer is yes! If not, say r of these women mention fewer than r available men, then these r women *plus* the r_0 women above have mentioned less than $r_0 + r$ men in total. This contradicts the property satisfied by these lists.

Thus we can invoke the induction hypothesis for the remaining m women and successfully have them marry as well.

This completes the proof by induction. ■

23.3. On Perfect Shuffling

To answer the first question, three perfect out-shuffles return a deck of eight cards to their original order! Surprisingly, just eight perfect out-shuffles do the same for a deck of 52 cards. Try it! (So, in what sense is a perfect shuffle "perfect"?)

To move the top card to any position in the deck, express that position as a binary number. (For example, 5 = 101 base 2.) Reading, from left to right, "1" as in-shuffle, "0" as out-shuffle, perform those maneuvers in that order. This places the card in the desired position! The same technique works for any size deck of cards.

This connection to binary numbers is a surprise. Here's why it works. Suppose N cards are in the deck, and the number $N - 1$ (corresponding to the position of the last card) has a binary representation r digits long. This means that $2^{r-1} \le N - 1 < 2^r$.

Subdivide the positions of cards into "blocks," the kth block being those positions with binary representations containing k digits:

$$0\ \overbrace{1}\ \overbrace{23}\ \overbrace{4567}\ \overbrace{89\cdots15}\ \overbrace{16\cdots31}\cdots\overbrace{2^{r-1}\cdots(N-1)}$$

We can obtain the digits in the kth block from those of the $(k - 1)$th block in two steps. First, multiplying them by 2 produces all the even numbers in the kth block. Second, multiplying by 2 and adding 1 produces the odd numbers in that block.

Multiplying by 2 corresponds to adding a 0 to the end of the $(k - 1)$-digit binary representation of a binary number to produce a k-digit one; multiplying by 2 and adding 1 corresponds to appending a 1. Thus, starting at the first block we can build up the binary representations of numbers in higher blocks by these two operations.

Now consider the operations of in- and out-shuffles. We see from the table that an in-shuffle takes a card from position t of the first half of the deck to position $2t + 1$, and an out-shuffle takes this card to position $2t$.

In-shuffle	Out-shuffle
$\mathbf{0} \to 4$	$\mathbf{0} \to \mathbf{0}$
$\mathbf{1} \to \mathbf{0}$	$\mathbf{1} \to 4$
$\mathbf{2} \to 5$	$\mathbf{2} \to \mathbf{1}$
$\mathbf{3} \to \mathbf{1}$	$\mathbf{3} \to 5$
$4 \to 6$	$4 \to \mathbf{2}$
$5 \to \mathbf{2}$	$5 \to 6$
$6 \to 7$	$6 \to \mathbf{3}$
$7 \to \mathbf{3}$	$7 \to 7$

These are precisely the operations of building up binary representations: start with the single entry in the first block, and move it up through the blocks via doubling, or doubling and adding one. Conveniently, performing "I" corresponds to appending a "1" and performing "O" appending "0"!

Acknowledgments and Further Reading

Martin Gardner writes extensively about mathematical card tricks in [Gard1], [Gard3], [Gard8], and [Gard13], for example. He also talks about perfect shuffles in [Gard17]. Also check out [Kola], [Pete1], [Pete3], and of course [Morr]. Card tricks and shuffling problems also appear in [Adle] and [Ball]. The curious piles puzzler appears as a game of solitaire in David Leep and Gerry Myerson's paper [Leep]. A discussion and proof of Hall's Marriage Theorem can be found in [Jaco] as well as [Ande].

24 Map Mechanics

24.1 Cartographer's Wisdom

This scheme, for example, does the trick.

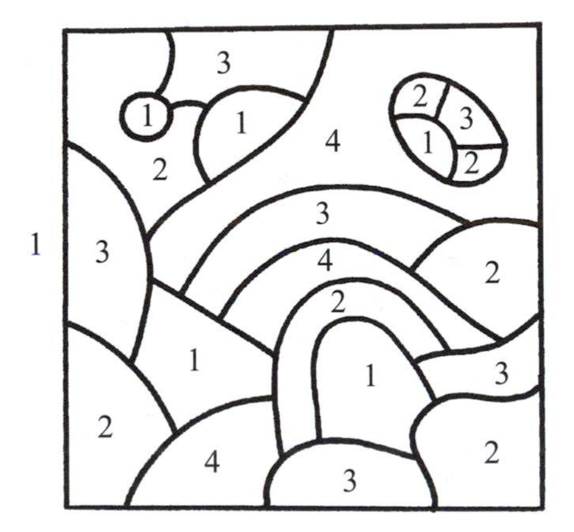

24.2 Simple Maps

All straight line and single curly line maps have the property that all interior vertices have *even degree*, that is, an even number of edges meet at all these points. (From our construction, you can see that any edge that enters an interior vertex must be followed by one that leaves it.) All even-degree maps are two-colorable. To see why, simply imagine teasing apart the interior vertices of such a map, as shown above. This converts the line maps into "nested islands." It is easy to see that a map of islands is two-colorable. This is enough to establish the claim.

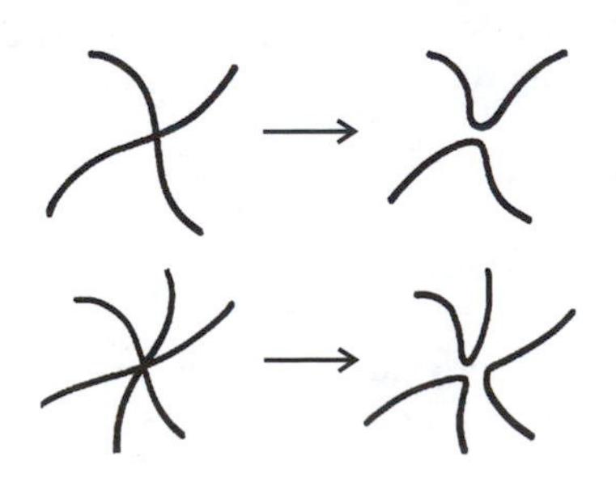

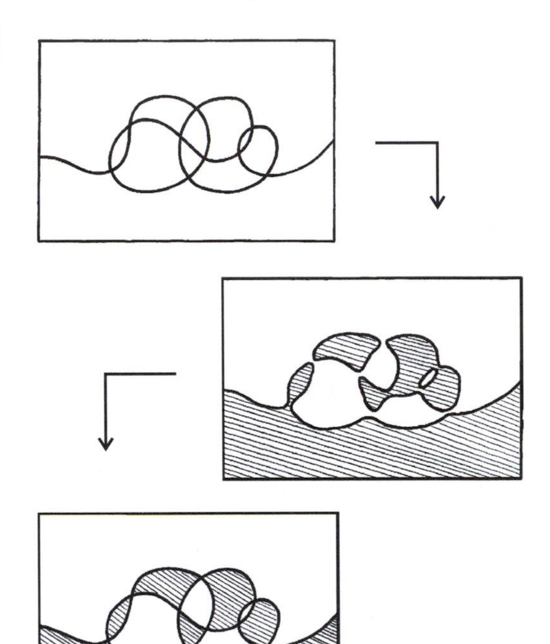

Challenge 1. A map is created from *several* curly lines wandering from one side of the page to the other. These curly lines intersect each other, and themselves, only at isolated points. Is the resulting map also two-colorable?

Challenge 2. A single curly line is drawn from one edge of the page to the other. It traces over entire lengths of itself. Will the resulting map necessarily be two-colorable?

Challenge 3. A piece of paper is folded onto itself multiple times to produce a flat origami design. When unfolded the creases in the paper outline a map of regions. Prove this map is necessarily two-colorable.

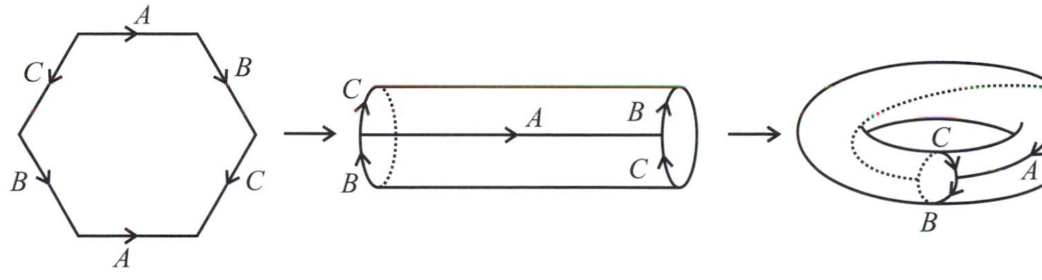

24.3 Toroidal Maps

One can also form a torus from a hexagon by gluing together opposite edges. See the figure above. (Sneaky!)

Here is a map that requires seven distinct colors to paint.

Challenge. What does this map look like on a fully formed torus?

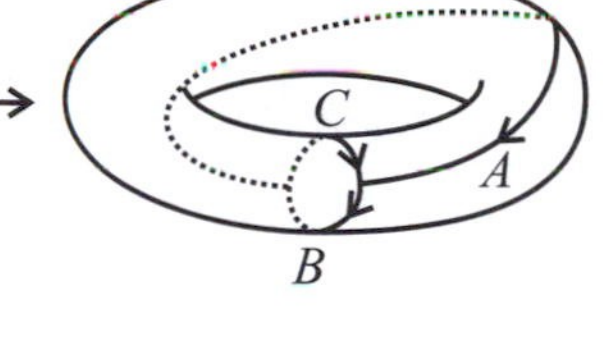

A Note on Toroidal Maps

It turns out that seven colors are sufficient to paint any map on a torus. Our aim here is to prove this. Surprisingly, it is not difficult!

We first make two assumptions about the maps given to us on a torus:

1. No region of the map "bounds a loop" in the sense indicated in the diagrams below. (If one does, draw in an extra edge to remove this difficulty. This does not alter the colorability of the diagram.)

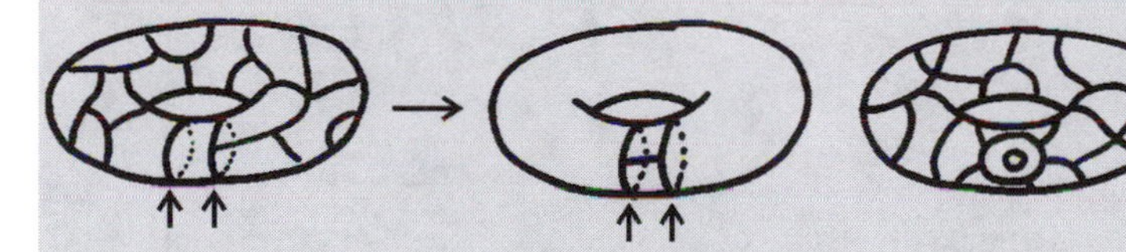

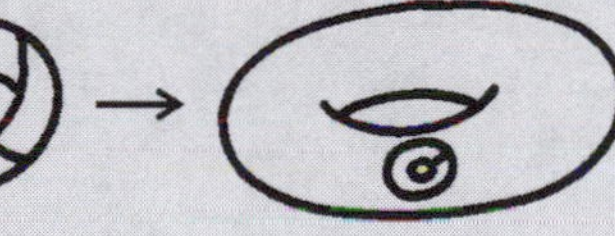

2. Precisely three edges meet at every vertex. (If not, draw in extra regions at the troublesome vertices. If this new map can be painted with just seven colors, then the original map certainly can be too, for "shrinking" each new region to a point returns the diagram back to the original map.)

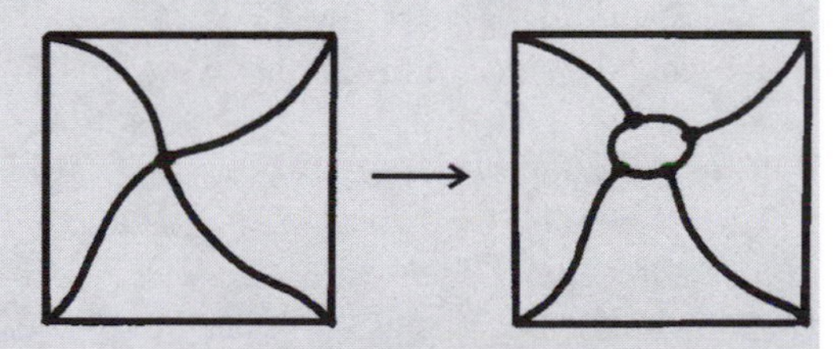

Recall from Chapter 14 that if r is the number of regions in a given toroidal map, e the number of edges, and v the number of vertices, then

$$v - e + r = 0$$

This is Euler's equation for a torus. We will use this equation to show that any map on a torus satisfying both assumptions possesses at least one region with six or fewer edges. This particular region will play a special role in proving our coloring theorem.

For a given map on a torus (satisfying both assumptions) let n_i, for each $i \in \mathbb{N}$, be the total number of regions in the map possessing precisely i edges. Thus

$$r = n_1 + n_2 + n_3 + \cdots.$$

The key trick now is to count the number of edges in two different ways. First, since each vertex has three edges coming to it, we need only count the number of vertices. Thus $3v = 2e$ (each edge is counted twice, once for each of the two vertices it connects). Second, we count the number of edges by counting the number of regions (noting each edge belongs to at most two regions). This yields

$$n_1 + 2n_2 + 3n_3 + \cdots \leq 2e.$$

Since $v - e + r = 0$ we obtain $2/3e - e + r = 0$, and so

$$e = 3r.$$

Thus

$$n_1 + 2n_2 + 3n_3 + \cdots \leq 2e = 6r = 6(n_1 + n_2 + n_3 + \cdots),$$

hence

$$0 \leq 5n_1 + 4n_2 + 3n_3 + 2n_4 + n_5 + 0 \cdot n_6 - n_7 - 2n_8 - \cdots.$$

If n_6 is zero, then it must be the case that at least one of the first five numbers n_1, n_2, n_3, n_4,or n_5 is positive. If n_6 is not zero, then n_6 is positive!

This establishes the existence of a region with six or fewer edges.

Challenge 1. Let a be the average number of edges the regions possess. (Thus a is some rational number.) Deduce $ar \leq 2e$. Use this to give an alternative proof to the existence of at least one region with six or fewer edges.

Theorem. *Seven colors suffice to paint any map on a torus.*

Proof. We will use an induction argument. Let $P(n)$ be the statement: "Any toroidal map with n regions can be painted with seven or fewer colors." Clearly $P(7)$ is true. Suppose $P(k)$ is true and we are given a map with k + 1 regions. Select a region with six or fewer edges. Remove one of its edges and meld together this special region and its neighbor into one single region. This leaves a map with k regions which, by the induction hypothesis, can be painted with seven or fewer colors.

Now put the excised edge back. Our special region is currently then the same color as one of its neighbors. As it has at most six neighbors there certainly is at least one color available to satisfactorily repaint it.

This completes the proof by induction. ■

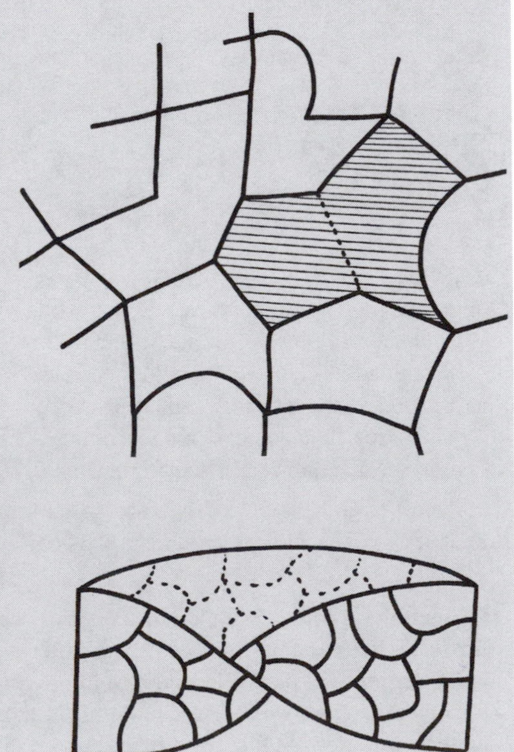

Challenge 2. How many colors suffice to color any map on a Möbius band? Assume the band is made of transparent material so that the color of any painted region shows through on both sides.

Acknowledgments and Further Reading

For some time, the *Four-Color Problem* was one of the most famous unsolved problems in mathematics: *Is it true that at most four colors suffice to paint any planar map*? (Perhaps experienced cartographers were never given sufficiently complicated diagrams.) After more than a century of failed attempts at proving it true or finding a counterexample, Appel and Haken finally showed in 1976 that the cartographers' hunch is indeed correct: Every planar map, no matter how complicated, can be satisfactorily painted with four or fewer colors. This was a monumental achievement. But their proof was not without controversy. Since the proof relied on 1200 hours of computer computation (checking nearly 2000 special arrangements of regions), the mathematics community found it difficult to accept this work with ease. Does such heavy reliance on a computer constitute a proof? Is this work comprehensible to the individual mind? More importantly, who can check that the computer correctly covered *all* possible special cases! The search is still on for a purely mathematical proof avoiding computer help altogether. (Though see [Robe]). To learn more about the Four-Color Theorem, its fascinating history, and its solution, have a look at [Ball], [Gard2], [Hake], [Hara], [May], and [Saat]. For map coloring on more complicated surfaces also see [Arno] and [Firb].

My thanks to Dr. Thomas Hull for alerting me to the origami map coloring problem. This little gem is is credited to Toshiyuki Meguro [Lang].

25 Weird Lotteries

25.1 Winning Cake

I have no solutions or optimal strategies to suggest for this problem. In practice this game seems to operate more on principles of human psychology than on rational analysis!

I once played a version of this game across the entire school campus, offering up to $100 of my own money to the player with the highest unique integer entry. (I promised to round up to the nearest penny in case of awkward fractions.) As expected, one person submitted a ridiculously large number "just because I want to win." And win they did—a whole penny! I knew all along my money would be safe.

There was one potential difficulty with this game that I didn't realize at the time. With a large number of people some contingent always likes to hand in ridiculously large numbers, and it is not always easy to tell which entrant is the winner. For example, of

$$\left(\left((100!)^{100!}\right)^{100!}!\right)!$$

and

$$99^{99^{99^{99^{9999}}}}+3$$

which number is the bigger? Luckily I could correctly determine the winner when I ran the game.

Taking it Further Answer. Everyone knows the remainder of the prize is going to be shared equally among the losers. So there is no point in submitting a number greater than the number of people in the room. If multiple entries are allowed and n people are in the room, then your best strategy would be to hand in one entry for each number from 1 to n. You are thus guaranteed $(1/n)$th of the cake, and you have a chance of winning more. Then again, everyone would reason the same way and do the same. Is it likely you would ever receive more than $1/n$ of the cake?

25.2 Unexpected Winner

The professor hasn't lied, but you, as a student do not know that yet!

Making the assumption that the professor has told the truth does indeed lead to a contradiction and so, logically, you must reject everything the professor said. It is as though the professor said nothing, in which case, John's name being read (or anyone else's for that matter) would be a surprise.

It is logically valid for the professor to read John's name after the two minutes are up, to have been entirely truthful in everything she said (she knew she was), and for it to be a complete surprise (logically) for all the students in the room!

Here's a similar case. Imagine I hand you a box and say, "In it you will find a plastic toy turtle." You have no way of knowing whether I am being truthful or not until you actually open it to find a plastic toy turtle. I, of course, know all along I am telling the truth, but at the outset you do not.

25.3 Winning Tootsie Rolls

These games are based on the famous Prisoner's Dilemma, in which two prisoners held in separate rooms must either confess to or deny involvement in a team crime. If both confess, each will be sentenced to two years of prison. But they'll get three years apiece if they both deny it (there is substantial evidence against this pair). However, if one denies and the other confesses, the denial carries just one year of jail time and the confession four! Not knowing how the other will respond, what should each prisoner do?

There are ways to analyze these games mathematically; [COMA] gives an elementary overview. But in practice human behavior comes to the fore. How did folks in your group respond? Does the size of the group affect general strategies? If you play with the same group of people multiple times, can you start predicting others' choices?

25.4 Buying Tootsie Rolls

By investing a fixed money amount from turn to turn, Clarence always gets the better deal on Tootsie Rolls®. Here's proof.

For $i \in \mathbb{N}$, let t_i denote the number of candies one penny can buy according to the ith roll of the die. After the first toss, Clarence has purchased t_1 Tootsie Rolls® for one penny and Denise has bought six for $6/t_1$ pennies. After n rolls of the die, Clarence has purchased $t_1 + t_2 + \cdots + t_n$ candies for n pennies, and Denise $6n$ for $6/t_1 + 6/t_2 + \cdots + 6/t_n$ pennies.

Clarence has paid an average cost of

$$\frac{n}{t_1+t_2+\cdots+t_n}$$

cents per Tootsie Roll® and Denise

$$\frac{\frac{6}{t_1}+\frac{6}{t_2}+\cdots+\frac{6}{t_n}}{6n}=\frac{\frac{1}{t_1}+\frac{1}{t_2}+\cdots+\frac{1}{t_n}}{n}$$

cents apiece.

I now claim that Denise's average cost is strictly greater than Clarence's average cost (assuming the values $t_1, t_2, \ldots, t_n$ are not all equal, otherwise the average costs are the same). It suffices to prove that

$$(t_1+t_2+\cdots+t_n)\left(\frac{1}{t_1}+\frac{1}{t_2}+\cdots+\frac{1}{t_n}\right)>n^2.$$

Look at the left-hand side (L.H.S.) of this inequality. It is

$$\begin{aligned}\text{L.H.S.} &= \left(\frac{t_1}{t_1}+\frac{t_2}{t_2}+\cdots+\frac{t_n}{t_n}\right)\\ &\quad+\left(\frac{t_1}{t_2}+\frac{t_2}{t_1}+\cdots+\frac{t_{n-1}}{t_n}+\frac{t_n}{t_{n-1}}\right)\\ &= n+\frac{n^2-n}{2} \text{ pairs of the form } \left(x+\frac{1}{x}\right).\end{aligned}$$

Now $x + 1/x \geq 2$ for any positive value of x with equality only for $x = 1$. (Check this!) So

$$\text{L.H.S.} > n+\frac{n^2-n}{2}\cdot 2 = n^2$$

(if not all the numbers t_i are equal). This completes the proof.

Comment. If the price of stock varies randomly from month to month, it is always better, in the long run, to invest a fixed amount of money in the stock, rather than purchase a fixed amount. However, if the price of the stock were to steadily decline, for example, this would no longer be the case!

Challenge. Clarence and Denise are each given three pennies. They agree to play as before until one runs out of pennies. If Denise at any stage cannot afford to buy six Tootsie Rolls® she purchases as many as her change will allow. Who, on average, ends up with the most Tootsie Rolls®?

Acknowledgments and Further Reading

The Unexpected Winner problem is a distilled version of the famous Unexpected Hanging paradox: A prisoner is sentenced to be hanged at dawn one day of the following week, Monday through Friday. The judge informs the prisoner that he will not know on what day he will be hanged until the morning of the event.

The prisoner, a savvy fellow, reasons as follows: *"I cannot be hanged on Friday. For having survived Monday through Thursday I would know my hanging day soon after dawn on Thursday, not on the morning of the day. With Friday ruled out, I can similarly argue that I will not be hanged on Thursday, for I will know this fact approximately 23 hours earlier, again not on the morning of the hanging day. By the same reasoning I can rule out Wednesday, Tuesday, and even Monday as the day of my execution. The judge's ruling cannot possibly be fulfilled—and therefore I cannot possibly be hanged!"* The prisoner is then surprised to find the executioner at his cell door Tuesday morning!

Martin Gardner writes beautifully about the origin of this paradox, its resolution, and other distilled versions in [Gard6], chapter 1. The Prisoner's Dilemma is another famous paradox causing much consternation and confusion to those who first encounter it. It is mentioned in [Paul] and [Vaki]. Game theoretic tools can be used to analyze it; see [COMA], [Binm], and [Stra]. A version of the dollar-averaging problem appears in [Chan].

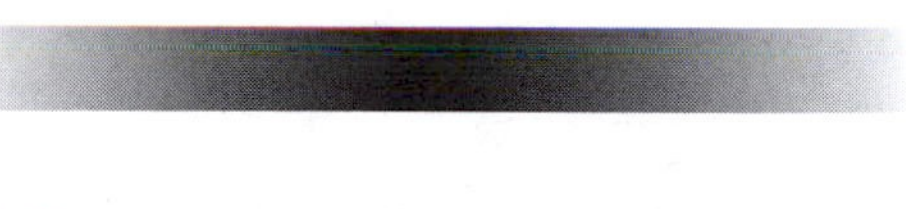

26 Flipped Out

26.1 A Real Cliff-Hanger

Number one-foot intervals along the expanse of land, placing 1 at the very edge of the cliff. Regard 0 as a position one foot into thin air: Dorothy's placement of doom!

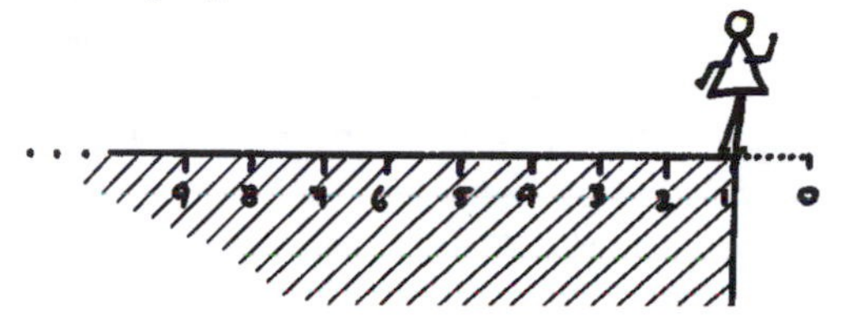

Let $p = P(1 \to 0)$ be the probability that, given Dorothy is standing at position 1, she will *eventually* visit position 0. More generally, define $P(i \to j)$ in the analogous way.

Walking from position 2 to position 1 requires precisely the same types of motions (shifted to the left by 1) as walking from position 1 to 0. We have therefore that

$$P(2 \to 1) = P(1 \to 0).$$

In fact,

$$P(i \to (i-1)) = P(1 \to 0) = p$$

for all $i \in \mathbb{N}$.

Let's now evaluate $p = P(1 \to 0)$. Suppose Dorothy is at position 1 (as at the start of the game). There is a 50% chance she will take a step forward to position 0 right away and a 50% chance she will take a step back. In the latter case, she will have to move from position 2 to position 0 later on to meet her doom. In order to do this, however, she must first move from 2 to 1 and then from 1 to 0. These observations lead to the following equations.

$$\begin{aligned} p &= P(1 \to 0) \\ &= P(\text{step forward}) \\ &\quad + P(\text{step back and reach 0 later on}) \\ &= \frac{1}{2} + \frac{1}{2} \cdot P(2 \to 0) \end{aligned}$$

$$= \frac{1}{2} + \frac{1}{2} \cdot P(2 \to 1 \text{ and } 1 \to 0)$$

$$= \frac{1}{2} + \frac{1}{2} \cdot P(2 \to 1) \cdot P(1 \to 0) = \frac{1}{2} + \frac{1}{2} \cdot p^2$$

Thus $p^2 - 2p + 1 = 0$, yielding $p = 1$. *With absolute certainty Dorothy will meet her doom!*

Challenge. If Dorothy is less likely to take a step forward, her chances of survival will increase. Suppose Dorothy carries with her a bag containing two white billiard balls and one red ball. She shakes the bag and pulls out a ball. If it is red she will take a step forward, a step back if it is white. She then replaces the ball to repeat this process indefinitely. Show that Dorothy's chances of survival are now 50%!

What are Dorothy's chances of survival if more white balls are placed into the bag?

26.2 Too Big a Difference

It is a surprise to learn that when tossing a coin n times, the number of heads and tails that appear differ, on average, by approximately $\sqrt{n/2}$. On the other hand, the average value of the square of this difference is, in the long run, precisely n. We establish the second claim below and leave the proof of the first claim as a note at the end of this section.

Let's introduce some variables. Set x_1 to be the value of the first toss:

$$x_1 = \begin{cases} +1 & \text{if first toss is heads,} \\ -1 & \text{if first toss is tails.} \end{cases}$$

Define $x_2, x_3, \ldots, x_n$ similarly. Then

$$S_n = |x_1 + x_2 + \cdots + x_n|$$

is the difference in the number of heads and tails that appear. Notice that

$$\begin{aligned} S_n^2 &= (x_1 + x_2 + \cdots + x_n)^2 \\ &= x_1^2 + x_2^2 + \cdots + x_n^2 \\ &\quad + 2x_1x_2 + 2x_1x_3 + \cdots + 2x_{n-1}x_n \\ &= 1 + 1 + \cdots + 1 + 2x_1x_2 + 2x_1x_3 + \cdots + 2x_{n-1}x_n \\ &= n + 2x_1x_2 + 2x_1x_3 + \cdots + 2x_{n-1}x_n. \end{aligned}$$

There are two possible values for each of the terms $2x_ix_j$ (i ≠ j),

$$2x_ix_j = \begin{cases} +2 & \text{if } x_i \text{ and } x_j \text{ have the same value,} \\ -2 & \text{if } x_i \text{ and } x_j \text{ have the opposite value.} \end{cases}$$

Each possibility is equally likely; 50% of the time $2x_ix_j$ will have value +2 and 50% of the time – 2. The average value of this term is thus 0. Consequently the average value of S_n^2 is

$$\text{Ave}\left(S_n^2\right) = n + 0 + 0 + \cdots + 0 = n.$$

as claimed.

Challenge 1. A child, playing a game, moves along a sidewalk according to the flips of a coin. If heads he takes one step forward; one step back if it lands tails. He repeats this procedure over and over again, thus performing a *one-dimensional random walk.*

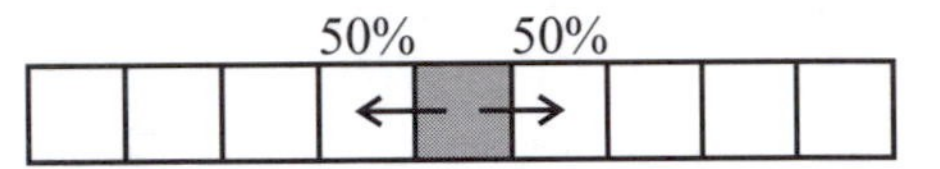

The claim above might lead us to believe that after n steps we can expect the child to be approximately $\sqrt{n/2}$ paces from his initial position, thus moving farther and farther from the origin as the number of steps increases. But the solution to section 26.1 shows that as soon as the child takes one step away from the origin he will, with absolute certainty (that is, with probability 1) again return to the origin.

Thus two competing tendencies are simultaneously operating on the child: to move both farther from the origin and closer to it. Which tendency wins out in the end? Can both simultaneously occur? How do we resolve this apparent paradox?

Comment. Molecules in a flowing stream both move along with the current and jolt back and forth in a random fashion across the current. Thus each individual molecule performs its own one-dimensional random walk in the direction orthogonal to the direction of flow.

Imagine a pollutant pouring steadily into the flowing water. Although we can never see the motion of an individual molecule, we can see the aggregate effect of many: the average behavior of the foreign molecules. They will concentrate along a path of the form $y = \sqrt{x/2}$, the formula for the average distance from the initial position. Thus a steady stream of pollutant in a flowing river spreads out in the shape of a parabola! (Have you ever had the unfortunate opportunity to observe this?)

Challenge 2. A child stands in an infinite grid of square paving stones. She will move from one stone to the next (in single vertical and horizontal steps) according to the roll of a four-sided die thus performing a *two-dimensional random walk*. A roll of 1 corresponds to one step north; 2 one step east; 3 one step south; and 4 one step west. (She could also use an ordinary six-sided die and ignore rolls of 5 and 6.)

If she played this game many times, performing n rolls in each case, how far from the origin would she end up on average? (Try performing the experiment!)

If she played this game for all eternity, she would, with probability 1, return to the origin infinitely many times. Show this!

Comment. The behavior of *three-dimensional random walks* is quite different. When taking a unit step up, down, left, right, back, or forth, with equal probability the chances of ever returning to your starting position again are now only about 34% [Doyl].

Challenge 3. How would you analyze a random walk on an infinite triangular grid of paving stones? How would you specify your motions with the roll of a die? Can anything be said mathematically about these walks?

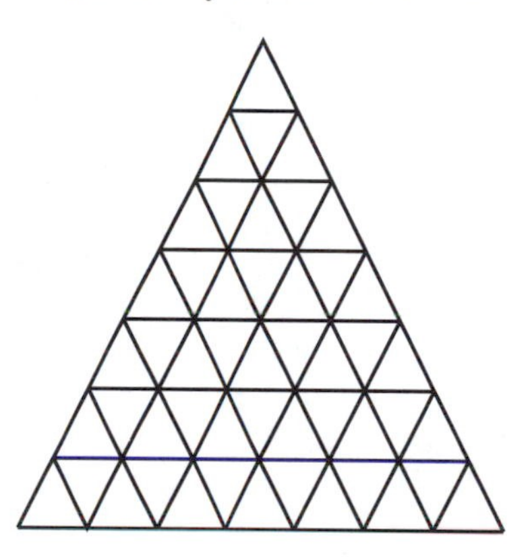

A Note on Ave(S_n)

Warning: This note is tricky. It assumes some familiarity with combinatorial coefficients, induction, and analysis of sequences of real numbers.

How can we compute Ave(S_n)? Let's begin with the case of $n = 3$ tosses. Eight equally likely possible outcomes exist for the values x_1, x_2, and x_3.

x_1	x_2	x_3	S_3
1	1	1	3
−1	1	1	1
1	−1	1	1
−1	−1	1	1
1	1	−1	1
−1	1	−1	1
1	−1	−1	1
−1	−1	−1	3

Thus

$$\text{Ave}(S_3) = \frac{\text{Sum}(S_3)}{8} = \frac{3+1+1+1+1+1+1+3}{8} = 1.5.$$

In the same way one can compute

$$\text{Ave}(S_1) = 1$$

$$\text{Ave}(S_2) = 1.$$

The table for the $n = 4$ experiment can be computed from the table for the $n = 3$ experiment. Double the number of rows in the table, repeat the values for x_1, x_2, and x_3 in the lower half, and insert a column of x_4 values—all "1" for the first eight rows, and all "– 1" for the remaining eight rows. Symmetry arguments allow us to compute the value of Ave(S_4) very quickly.

For the rows with $x_4 = 1$, if the majority of values among x_1, x_2, x_3 are positive, then the value of S_4 is one more than the corresponding value of S_3 in the $n = 3$ table. If the majority of values among x_1, x_2, x_3 are negative, then the value of S_4 is one less than the corresponding value of S_3. Each scenario occurs equally often and so the sum of the values of S_4 within just the first eight rows in the $n = 4$ table is just Sum(S_3), the sum of the values S_3 in the $n = 3$ table. Similarly, the sum of the values in the last eight rows is also equal to Sum(S_3). Thus

$$\text{Sum}(S_4) = 2 \cdot \text{Sum}(S_3).$$

As there are twice as many rows in the table for $n = 4$ as there are in the table for $n = 3$, we have

$$\text{Ave}(S_4) = \text{Ave}(S_3) = 1.5.$$

This argument shows in general that

$$\text{Ave}(S_{n+1}) = \text{Ave}(S_n)$$

whenever n is odd.

The same argument can be applied to the case where n is even. One more scenario should be considered: the existence of an equal number of positive and negative terms among x_1, $x_2, \ldots, x_n$. These rows have a corresponding value $S_n = 0$ and have no role in the calculation of Sum(S_n). However, in appending $x_{n+1} = 1$ or $x_{n+1} = -1$ to these rows, we obtain a value $S_{n+1} = 1$, and these values will have an effect in the computation of Sum(S_{n+1}). (The effects of appending an x_{n+1} column to the remaining rows cancel, just as they did for the case n odd.)

There are $\binom{n}{n/2}$ rows with this property: the number of ways $n/2$ terms can be selected to be negative among a row of n terms. Each of these rows will be appended with $x_{n+1} = 1$ and then with $x_{n+1} = -1$. It follows

$$\text{Sum}(S_{n+1}) = 2 \cdot \text{Sum}(S_n) + 2 \cdot \binom{n}{\frac{n}{2}}.$$

The table for S_n has 2^n rows, and the table for S_{n+1} has 2^{n+1} rows. Thus:

$$\text{Ave}(S_{n+1}) = \text{Ave}(S_n) + \frac{1}{2^n}\binom{n}{\frac{n}{2}}$$

whenever n is even. These recursive relations allow us to compute the values Ave(S_n). For example,

n	1	2	3	4	5	6	7	8	9	10
Ave(S_n)	1	1	1.5	1.5	1.875	1.875	2.188	2.188	2.461	2.461

Our next aim is to show that these values can be reasonably approximated by the formula $\sqrt{\frac{n}{2}}$, at least for large values of n.

As the values of Ave(S_n) come in pairs, it is easier to work with every second term of the sequence. Let A_n be the value of the nth pair of this sequence. Then A_n satisfies the recursion formula

$$A_1 = 1$$
$$A_{n+1} = A_n + \frac{1}{4^n} \cdot \binom{2n}{n}.$$

It follows that

$$A_{n+1} = 1 + \sum_{k=1}^{n} \frac{1}{4^k} \cdot \binom{2k}{k}.$$

Lemma 1. $\frac{1}{4^n} \cdot \binom{2n}{n} \geq \frac{1}{2} \cdot \frac{1}{\sqrt{n}}$ for all n.

Proof. Use an induction argument. The claim is certainly true for $n = 1$. Suppose it is true for a value $n = k$. Then

$$\frac{1}{4^{k+1}} \cdot \binom{2k+2}{k+1} = \frac{1}{2} \cdot \frac{2k+1}{k+1} \cdot \frac{1}{4^k} \cdot \binom{2k}{k} \geq \frac{1}{2} \cdot \frac{2k+1}{k+1} \cdot \frac{1}{2\sqrt{k}}$$

by the induction hypothesis. This latter quantity can easily be shown to be $\geq 1/2\sqrt{k+1}$ by squaring each quantity and doing the algebra. ■

Result 1. $A_n \geq \sqrt{n}$ for all n.

Proof. This is certainly true for $n = 1$ and 2. Notice

$$A_{n+1} = A_n + \frac{1}{4^n} \cdot \binom{2n}{n} \geq \sqrt{n} + \frac{1}{\sqrt{2}} \cdot \frac{1}{\sqrt{n}} \geq \sqrt{n+1}.$$

The first inequality follows by an induction hypothesis, and the second by algebra. ■

Lemma 2. $\frac{1}{4^n} \cdot \binom{2n}{n} \leq \frac{1}{\sqrt{\pi}} \cdot \frac{1}{\sqrt{n}}$ for all n.

Proof. The result is true for $n = 1$, 2, and 3.

James Stirling, a friend of Newton, came close to writing in his work a remarkable formula (one that now bears his name), namely:

$$\lim_{n \to \infty} \frac{n!}{\sqrt{2\pi n} n^n e^{-n}} = 1$$

(see [Simm] for a proof of this result). This means the function $\sqrt{2\pi n} \cdot n^n \cdot e^{-n}$ is a very good approximation for $n!$; the relative difference between the two formulae goes to zero as n becomes large. Thus

$$\frac{1}{4^n} \cdot \binom{2n}{n} = \frac{1}{4^n} \cdot \frac{(2n)!}{(n!)^2}$$

can be well approximated by

$$\frac{1}{4^n} \cdot \frac{\sqrt{4\pi n} \cdot (2n)^{2n} \cdot e^{-2n}}{2\pi n \cdot n^{2n} \cdot e^{-2n}} = \frac{1}{\sqrt{\pi}} \cdot \frac{1}{\sqrt{n}}.$$

Precisely,

$$\lim_{n\to\infty} \frac{\frac{1}{4^n}\cdot\binom{2n}{n}}{\frac{1}{\sqrt{\pi}}\cdot\frac{1}{\sqrt{n}}} = 1.$$

One can show that $\left\{\frac{1}{4^n}\cdot\binom{2n}{n}\cdot\sqrt{\pi}\cdot\sqrt{n}\right\}$ is an increasing sequence of values (algebra!). It approaches the value 1 as n grows, and consequently is bounded above by the value 1. Thus $\frac{1}{4^n}\cdot\binom{2n}{n}\cdot\sqrt{\pi}\cdot\sqrt{n} \le 1$ for all values of n. ■

Result 2. $A_n \le \frac{2}{\sqrt{\pi}}\cdot\sqrt{n}+1$ for all n.

Proof. This is true for $n = 1$. Now

$$A_n = 1 + \sum_{k=1}^{n-1} \frac{1}{4^k}\cdot\binom{2k}{k} \le 1 + \frac{1}{\sqrt{\pi}}\sum_{k=1}^{n-1}\frac{1}{\sqrt{k}}$$

by lemma 2. We need to show this quantity is $\le \frac{2}{\sqrt{\pi}}\cdot\sqrt{n}+1$. This is equivalent to showing

$$1 + \frac{1}{\sqrt{2}} + \cdots + \frac{1}{\sqrt{n-1}} \le 2\sqrt{n}.$$

This is certainly true for $n = 2$ and can be established for all n by an induction argument. ■

Thus we have shown

$$\sqrt{n} \le A_n \le \frac{2}{\sqrt{\pi}}\cdot\sqrt{n}+1$$

for all n. Now $\frac{2}{\sqrt{\pi}} \approx 1.1284$. Thus $\sqrt{n}$ is a reasonable approximation for A_n. (Well, at least in the sense that A_n is bounded by the square root function.) As A_n represents a pair of terms in the sequence $\{\text{Ave}(S_n)\}$, it follows that $\sqrt{\frac{n}{2}}$ is a reasonable approximation for the value $\text{Ave}(S_n)$. This establishes the first claim made in our solution to section 26.2.

26.3 A Surprise

A balanced penny is more likely to fall heads up rather than tails up, yet a spinning penny is more likely to land tails up rather than heads! (There seems to be no such bias for other American coins.) Curiously, American pennies are manufactured with a marked bias. Be careful next time when a friend wants to flip a coin!

Acknowledgments and Further Reading

Peter Doyle and J. Laurie Snell's marvellous book [Doyl] offers a wonderful account of the theory of random walks and their analysis. My thanks to Dr. Fozia Qazi for alerting me to the cliff-hanger problem. The bias in American pennies has been noted by many people; for example, [Pete5].

27 Parts That Do Not Add Up to Their Whole

27.1 A Fibonacci Mismatch

If you look carefully at the rectangular arrangement of the 8 × 8 square you will notice that

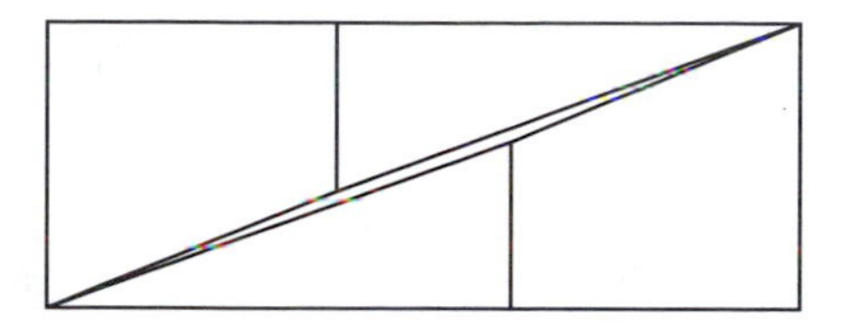

the pieces don't quite line up correctly. Usually we blame such discrepancies on cutting errors, but in this case the errors are inherent to the problem. A gap in the middle of the rectangle accounts for the missing unit area.

Taking it Further. One can prove, via induction, that the Fibonacci numbers satisfy:

$$F_n^2 = F_{n-1} \cdot F_{n+1} \pm 1.$$

(To establish $F_{n+1}^2 = F_n \cdot F_{n+2}$ ml from $F_n^2 = F_{n-1} \cdot F_{n+1} \pm 1$ use $F_{n+2} = F_{n+1} + F_n$.)

This shows that a Fibonacci square and the corresponding Fibonacci rectangle always differ in area by one unit. For large-sized squares, the difference is difficult to detect. For small squares, such as 3 × 3 and 5 × 5 squares, the difference is quite pronounced. (Try it!)

There are many examples of such cutting tricks, all relying on the imperceptible overlapping of edges. Here's another one. Make two enlarged copies of the triangle on the left. Rearrange the pieces of one copy as shown on the right. Place these pieces on top of the other photocopy and verify that, despite the appearance of a missing rectangle, the size of the second triangle is exactly the same as the first!

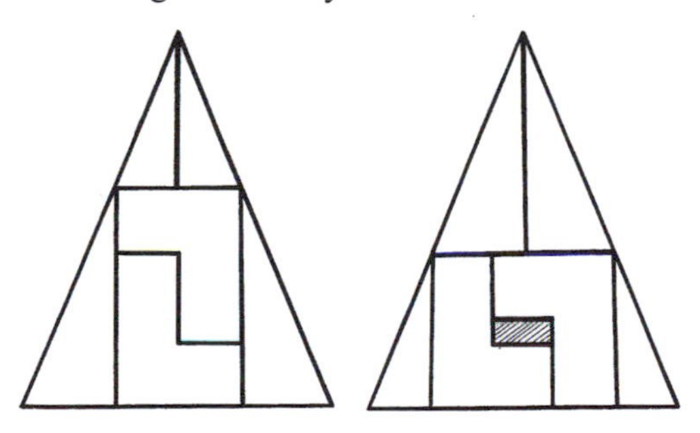

27.2 Cake Please

Every proportional cake-sharing scheme between two people is automatically envy-free. (Can you see why?) In particular, the "you cut, I choose" and the parallel knife cut methods for sharing cake between two people are both proportional and hence envy-free. They are specific cases of one general scheme which we can summarize as follows.

1. Player 1 cuts the cake into two pieces, P_1 and P_2, which he believes are of equal size.
2. Player 2 takes what she believes to be the larger of these two pieces, say P_2, and trims off any excess, T, to create two pieces P_1 and $P_2 - T$ of equal size in her estimation and some excess trim T. (Player 2 could also believe that P_1 and P_2 are of the same size and not trim anything at all; in which case, the process stops.)
3. Player 1 takes piece P_1 and player 2 piece $P_2 - T$.

In "you cut, I choose," player 2 also takes all the trim. In the parallel knife cut"scheme, the trim is shared in some arbitrary manner between the two players. But let's be clever about sharing the trim. This will lead to a way of devising an envy-free sharing procedure between three players.

At this stage player 1 believes $P_1 > P_2 - T$ and player 2 believes $P_1 = P_2 - T$.

4. Player 2 divides the trim into two pieces T_1 and T_2 which she believes are of equal size.
5. Player 1 takes the piece of trim he believes to be larger, say T_1. He keeps P_1 and T_1. (If player 1 believes $T_1 = T_2$, he can just take either one.)
6. Player 2 keeps $P_2 - T$ and T_1.

As a result of this scheme player 1 believes $P_1 + T_1 > (P_2 - T) + T_2$ and so will not envy player 2. Player 2 believes $(P_1 - T) + T_2 = P_1 + T_1$ and will not envy player 1.

Here is the generalization to an envy-free procedure among three players.

1. Player 1 cuts the cake into three pieces, P_1, P_2, and P_3 he considers to be of equal size.
2. Player 2 selects a piece, say P_2, she believes to be the largest (or one of the largest). She trims it so that there is now definitely, in

her mind, a tie for the largest. We now have three pieces, P_1, $P_2 - T$, and P_3. The trimmings T are set aside.

3. Player 3 chooses the piece she considers to be the largest (or one of the largest).
4. Player 2 chooses from the remaining two pieces the one she believes to be larger with the proviso that she must choose $P_2 - T$ if it is still available.
5. Player 1 takes the remaining piece.

Player 3 got to choose among the three pieces, so at this stage she certainly believes she has the largest piece. Player 2 created a two-way tie for largest and certainly, after player 3 has made her choice, one of these largest pieces is still available for her. Player 1 does not end up with piece $P_2 - T$ because of the proviso in step 4, so he believes his piece to be no larger than any other piece (by step 1). So far we are in an envy-free situation—but there are still the trimmings T to contend with.

Suppose it was player 3 who took piece $P_2 - T$ (if it was player 2, modify the remainder of the procedure below appropriately.)

6. Player 2 cuts the trim T into three pieces she considers equal in size.
7. Player 3 chooses the one he believes to be the largest.
8. Player 1 chooses from the remaining two pieces the one he believes to be the larger.
9. Player 2 takes the remaining piece of trim.

Player 3 already believes she has the largest piece of cake and then gets to select the largest piece of trim. She certainly envies no one! Player 1 does not envy player 2 because he is choosing trim ahead of her; and he does not envy player 3 because adding a portion of trim to $P_2 - T$ produces a quantity still smaller than P_2 and hence smaller than the size of his own piece (recall step 1). Finally, player 2 already believes she has the largest piece of cake from the first part of the scheme, and she made all three pieces of T the same size.

27.3 Sharing Indivisible Goods

The person who bid the most for a particular item should receive that item. Thus Sheryl receives the dacquoise, Neal the torte and the pavlova, and Janice the brownie. Neal thus receives items worth 154 Tootsie Rolls® more than his estimation of a third of the share; Sheryl an excess of 28 Tootsie Rolls®; and Janice suffers a deficit of 87 Tootsie Rolls®.

	Neal	Janice	Sheryl
Dacquoise	105	120	**132**
Torte	**90**	80	64
Pavlova	**196**	75	112
Brownie	5	**7**	4
Total	396	282	312
One-Third Share	**132**	**94**	**104**
Goods Received	286	7	132
Excess	+132	– 87	+28

Neal and Sheryl should place 154 and 28 Tootsie Rolls®, respectively, into the center of the table, and Janice should take 87 from this pile. This puts everyone in a balanced position, each currently possessing precisely a one-third share of the goods (in his or her estimation) or their Tootsie Roll® equivalents. It also creates a pot of 95 extra Tootsie Rolls®. Neal, Janice, and Sheryl can now divvy up this pot between them and each be 31⅔ candies ahead! (Although they were averse to slicing the desserts, they weren't to slicing Tootsie Rolls®!)

Acknowledgments and Further Reading

The missing-area problems are classic teasers. Martin Gardner discusses them at length in his wonderful book [Gard1]. The Fibonacci Mismatch problem also appears in [Vaki] and Curry's triangle teaser in [Bunc].

The envy-free cake-sharing method described in section 27.2 was discovered independently by John L. Selfridge and John H. Conway in the 1960s. With modification it extends to four or more players. A very accessible account of this method and other schemes for sharing cake and indivisible goods can be found in [COMA], chapter 13. See also [Bram1], [Bram2], [Dubi] (advanced), and [Hons3]. For an elegant envy-free cake-sharing procedure based on Sperner's Lemma, see Francis Edward Su's wonderful article [Su].

28 Making the Sacrifice

28.1 The Josephus Flavius Story

Beginning with $W(1) = 1$ we see that the value of $W(n)$ is 3 more than the previous value, subtracting n if $W(n-1) + 3$ is too big. For example, $W(8) = W(7) + 3$ and $W(14) = W(13) + 3 - 14$. Moving n (or any multiple of n) places corresponds to counting around the full circle a number of times. Subtracting this value from a formula does not affect the computation of the winner's placement.

This pattern persists for all other games as well. For instance, counting every ninth person produces the table:

n	1	2	3	4	5	6	7	8	9	10	11	12	13	14	15
$W(n)$	1	2	2	3	2	5	7	8	8	7	5	2	11	6	15

This observation makes it easy to write out a table for any game and quickly predict who will be the winner. It is easy to see why it works. After the first person is eliminated in an n-person game, counting off every mth participant, leaves an $(n-1)$-person game starting at a position shifted m places along. Thus $W(n) = W(n-1) + m$, with adjustments to account for wrapping around the full circle.

Challenge 1. Is it possible to predict who will be the second place winner in any game of n people counting every mth place?

This game is based on the story of Josephus Flavius, a Jewish military leader fighting against the Romans in the first century A.D. In the spring of 67, Josephus's men were under siege in the town of Jotapata and with little chance of survival. Rather than submit their fates to the hands of the Romans, Josephus' men decided to take refuge in a cave and commit group suicide. Josephus, it is said, did not agree with this idea, but offered no open opposition to it. Instead, he suggested they maintain order by sitting in a circle and humanely killing every third man until only one person was left, who would commit suicide. Josephus apparently positioned himself so that he was the sole survivor of this scheme. Instead of committing suicide he surrendered to the Romans.

Challenge 2. Ten students, five men and five women, attend a mathematics club. However there is a terrible shortage of snacks and five people must leave. The participants decide to sit in a circle and count off every 11th player, who must then leave. They do this until only five students are left. The women cleverly arrange themselves about the circle as shown and begin the count, clockwise, at the position indicated. What happens?

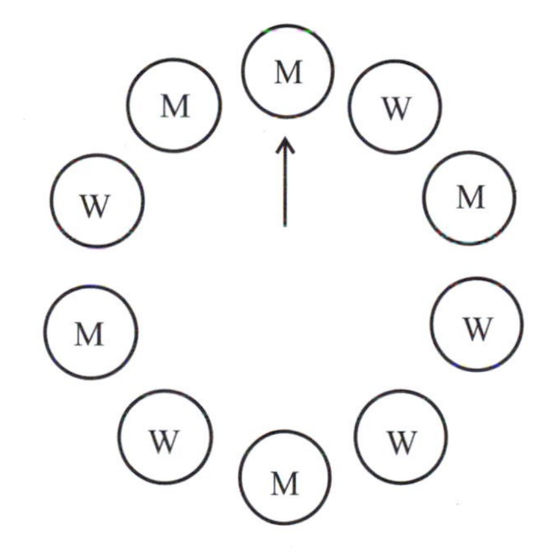

The men, seeing how the women have arranged themselves, quickly suggest counting clockwise every 29 places, starting two places to the left! Now, what happens?

28.2 Soldiers in the Desert

It is a surprise to learn that it is absolutely impossible to push a penny five (or more) lines into the desert! No matter what ingenious configurations you devise behind the border line, nothing will push a penny five places in. We

								144							
2	3	5	8	13	21	34	55	89	55	34	21	13	8	5	3
1	2	3	5	8	13	21	34	55	34	21	13	8	5	3	2
1	1	2	3	5	8	13	21	34	21	13	8	5	3	2	1
	1	1	2	3	5	8	13	21	13	8	5	3	2	1	1
		1	1	2	3	5	8	13	8	5	3	2	1	1	
			1	1	2	3	5	8	5	3	2	1	1		
				1	1	2	3	5	3	2	1	1			
					1	1	2	3	2	1	1				
						1	1	2	1	1					
							1	1	1						
								1							

can prove this amazing fact with a clever use of the Fibonacci numbers. These are the numbers arising in the sequence 1, 1, 2, 3, 5, 8, 13, 21, 34, 55,... defined recursively by $F_n = F_{n-1} + F_{n-2}$ with $F_1 = F_2 = 1$.

Suppose there is a starting configuration of coins below the border line that, after n checker jumps, leaves a coin five places into the desert. Label the grid with Fibonacci numbers as shown above, with the place on the fifth row where the final coin lands labelled 144. (*Note*: We are assuming that the initial layout of coins fits inside the triangle of numbers below the line. If not, use a larger triangle of numbers. The proof will be analogous to this one.)

Let

W_0 = the sum of all the numbers covered by coins in the initial arrangement,

W_1 = the sum of all numbers covered by coins after one jump has occurred,

W_2 = the sum after two jumps have occurred,

and so on. Notice that a jump removes two coins and places one adjacent to the two removed. This has two possible effects on a sum W_i: Either the sum of numbers covered by coins decreases

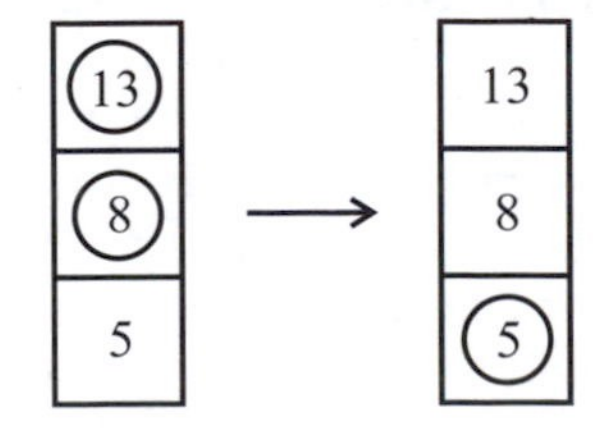

or it remains unchanged. (Recall the relation that the Fibonacci numbers satisfy.) Thus we have

$$W_0 \geq W_1 \geq \cdots \geq W_n.$$

Now, the final configuration covers the number 144 at least. Thus

$$W_n \geq 144.$$

The initial configuration, however, covers only those numbers below the border line. Thus

$$\begin{aligned} W_0 \leq\ & 1+1+2+3+5+8+13+8+\cdots+2+1+1 \\ & +1+1+2+\cdots+2+1+1 \\ & +1+1+\cdots+1+1 \\ & +1+1+2+3+2+1+1 \\ & +1+1+2+1+1 \\ & +1+1+1 \\ & +1 \\ =\ & 125. \end{aligned}$$

This contradiction shows that no coin could ever reach five lines into the desert!

Comment. We have relied on a property that the sum of a triangle of Fibonacci numbers centered about F_n is always less than F_{n+5}. This turns out to be true in general and is discussed in Martin Aigner's article [Aign]. This activity is based on his discussion.

28.3 Democratic Pirates

The key to analyzing this game is to start with a few pirates and work up to a crew of ten. Now, if there is just one pirate he will keep all ten coins and happily vote for himself to do so! If there are two pirates, the second pirate can keep all ten coins and vote for this distribution scheme. He thus garners 50% of the votes and succeeds.

If there are three pirates, the third will reason that if he dies, the captain gets no coins, given the two-pirate scheme. Thus he can buy the captain's vote simply by offering him a single coin. He should propose one coin to pirate 1, no coins to pirate 2, and nine to himself. He would win two thirds of the votes.

Continuing this way, we can build up a table of optimal distribution schemes. We see that the cabin boy can land himself survival and six gold coins!

Number of Pirates	Distribution									
1	10									
2	0	10								
3	1	0	9							
4	0	1	0	9						
5	1	0	1	0	8					
6	0	1	0	1	0	8				
7	1	0	1	0	1	0	7			
8	0	1	0	1	0	1	0	7		
9	1	0	1	0	1	0	1	0	6	
10	0	1	0	1	0	1	0	1	0	6

Challenge. The ten democratic pirates, drifting at sea with no food (but plenty of water), decide to resort to cannibalism. They will vote on whether to eat or spare each crewman, starting with the cabin boy and then moving up the ranks. A simple majority vote to "eat" will ensure a crewman's demise. They will stop taking votes as soon as someone is spared. Given this scheme, will the cabin boy survive?

Acknowledgments and Further Reading

The story of Josephus Flavius and other sacrificial games can be found in W. W. Rouse Ball and H. S. M. Coxeter's fabulous book [Ball]. For further historical information see Jona Lendering's ancient history website [Lend]. The Soldiers in the Desert game also appears in E. Berlekamp, J. H. Conway and R. Guy's incredible text [Berl2].

I cannot track down the source of the Democratic Pirates problem. It was making the rounds through the MIT mathematics department a few years back. The cannibalistic version has been explored (but not published) by my colleagues, Dr. Charles Adler and Dr. Raymond Robb, and myself.

29 Problems in Parity

29.1 Magic Triangles

I made sure you commenced the game on a black cell. Every journey of an even number of steps takes you to a cell of the same color; an odd number of steps takes you to one of opposite color. The first instruction asked you to move to a white cell (11 steps) and thus I knew the multiples of three could safely be removed after this maneuver. I followed the same trick for the remaining seven instructions, whittling the board down to a single number in the process—number 23!

Challenge. Invent your own magic tricks of this type for different triangular grids, square grids, and even cubical lattices. Can a similar puzzle be devised with a grid of hexagons arranged in a honeycomb pattern?

29.2 Walking a Loop

Because each crossing point is visited twice, each letter appears twice in a sequence describing a journey. (We ignore the final letter listed, which is already mentioned once as a terminal point of the walk.) Gauss observed, furthermore, that curves of this type have the property that each crossing point appears in a sequence once in an odd position and once in an even position. Putting it another way, an even number of letters always separates any two crossing points in the list.

Thus, writing the list alternately above and below a horizontal line makes each crossing point appear once above the line and once below the line. If two neighboring crossing points are interchanged, each appears twice on one side of the line and thus can be identified immediately. (Try it! As your friend reads out her list, transcribe it on your own piece of paper using a separating line.)

We establish Gauss's observation in the following note.

A Note on Closed Self-Intersecting Curves

Consider a closed self-intersecting curve C that passes through each of its crossing points only twice, and consider an arbitrary crossing point P on the curve. We will assume that all crossing points on the curve are legitimate, in that they don't represent places where two sections simply touch tangentially. If we start at position P and trace along C we will eventually return to P. Our aim is to prove that an even number of crossing points (counted with multiplicity) are encountered along the way.

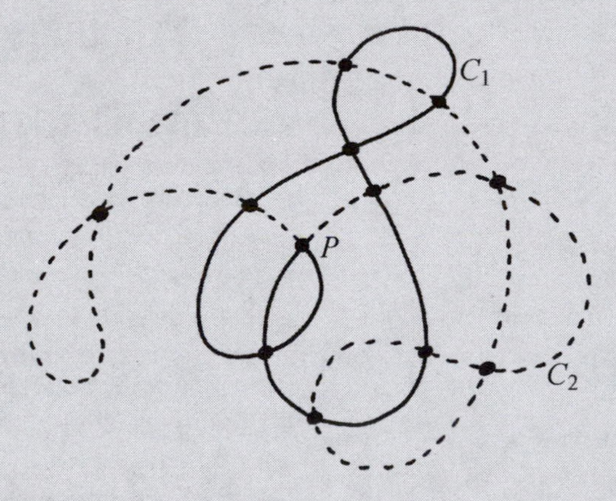

When we first return to P, our finger has traversed only part of the curve. Call this portion C_1. It too forms a closed self-intersecting curve (with a sharp angle at P). The remaining portion of the curve, C_2, also forms its own closed self-intersecting loop. Let's deform our picture slightly by teasing apart these two sub-loops, C_1 and C_2, at the point P.

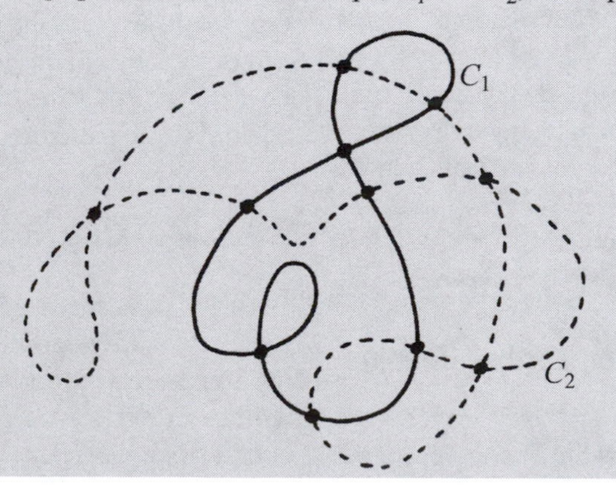

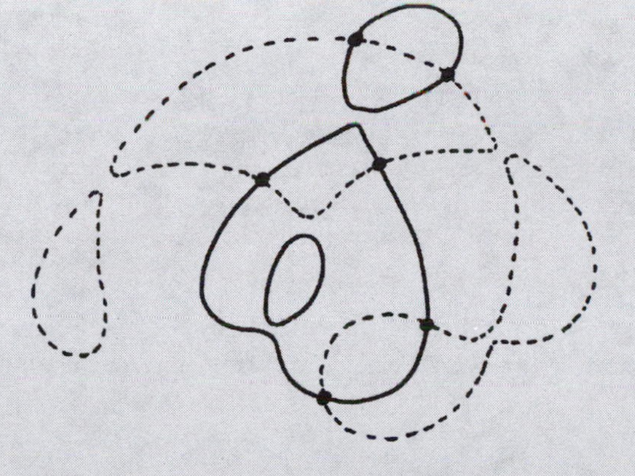

The loops C_1 and C_2 may themselves possess crossing points. Because each such crossing point will be listed twice when traversing the curve C_1 (and C_2 for that matter) from point P back to point P, they each contribute a count of two to our list. Our interest then should focus on the number of times C_1 and C_2 intersect.

"Dissolve" any crossing point that C_1or C_2 possesses in the manner indicated in the diagram above. (*Note:* There are two ways to "dissolve" any particular crossing point. Your choices can be arbitrary.) This maneuver decomposes each of C_1 and C_2 into an array of disjoint (but possibly nested) islands, that is, a disjoint collection of closed loops, none of which are self-intersecting. Let's call these disparate pieces circles. Like an ordinary circle, each possesses a well-defined interior and a well-defined exterior.

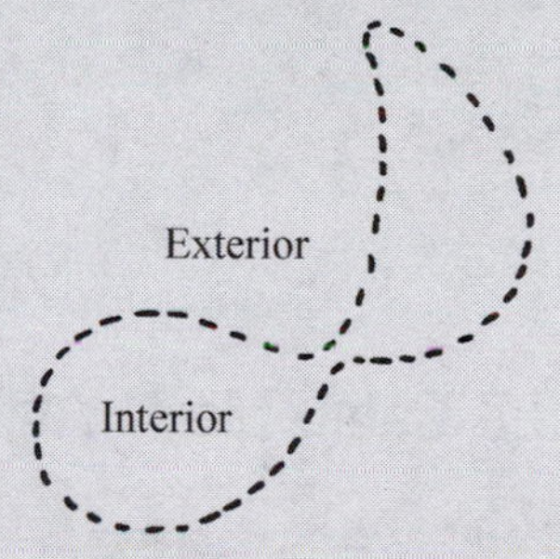

Two circles can intersect only in an even number of points (or not at all). To see why, imagine tracing along one circle with your finger. You must alternately enter and leave the interior of the other circle, thus forcing the number of points of intersection to be even. (Recall, we assumed that the curve C never just touched itself at any given point.) Thus the curves C_1 and C_2 must intersect an even number of times, so the total number of crossing points encountered in traversing C_1 to and from the point P, including its own crossing points (counted with multiplicity) must be even. This establishes Gauss's result.

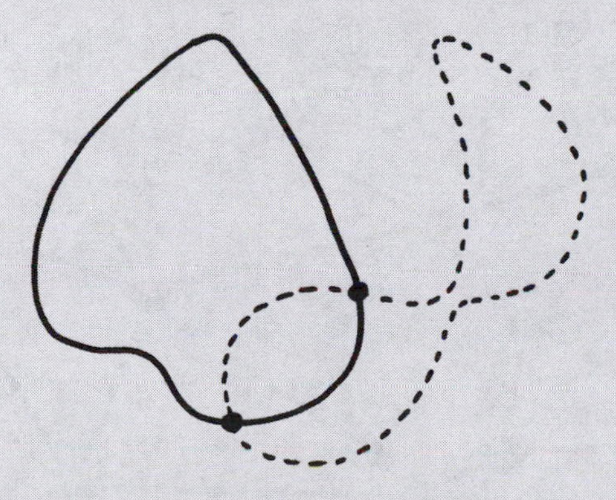

Challenge 1. This proof relied on **Jordan's Curve Theorem,** which states that any circle drawn on a plane divides that plane into two regions: an interior and an exterior. As we saw in

Section 14.2, this theorem is not true for all curves drawn on a torus. As a consequence it is possible to have two toroidal circles intersect each other an odd number of times. Does this mean Gauss's observation also fails for closed self-intersecting curves drawn on a torus?

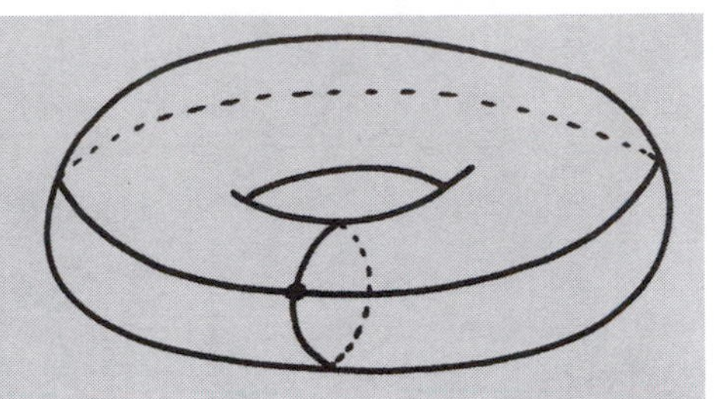

Challenge 2. A closed self-intersecting curve drawn on a plane passes through each of its crossing points possibly more than twice. Can Gauss's trick be modified to work in this general situation?

29.3 Catch Me If You Can

If Port Moresby is removed from the diagram, the remaining locations can be colored black or white in a pseudo-checkerboard scheme (see the figure below). No matter how Craig moves, Joy will always land on a square opposite in color to his at the end of each of her turns, so she will never catch him. Her best bet is to head to Port Moresby and change the parity of her predicament. Once this is done, she will usually corner Craig within the allotted number of moves. (Armed with this knowledge, you will find this game an excellent teaser to try out on unsuspecting friends!)

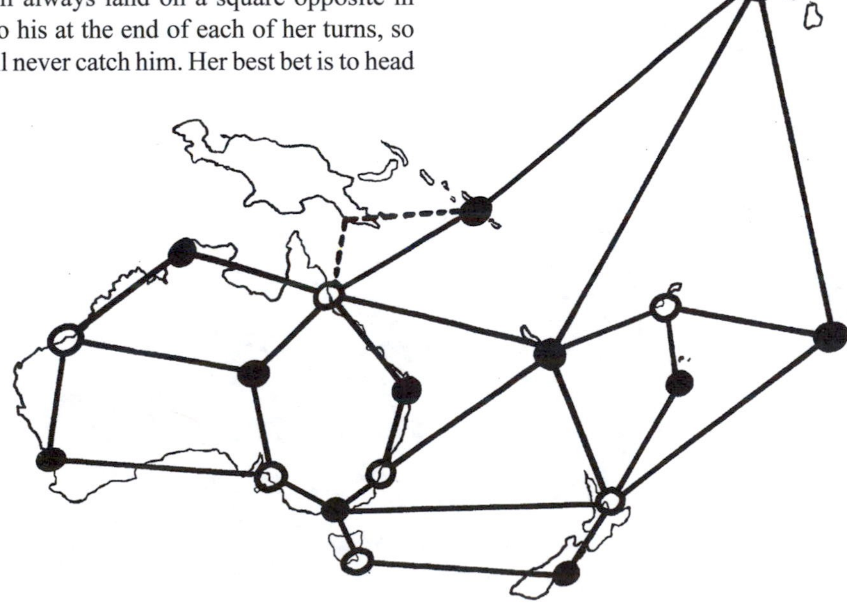

29.4 A Game of Solitaire

(*Warning:* An induction argument!) The game can be won if, and only if, an odd number of heads is showing initially. You may have observed this when playing with a small number of coins. It is certainly the case with one coin—you need a "head" to even make a move! You will have also noticed that removing an interior coin from the game breaks the line into two segments of shorter length. Each segment represents a game unto itself. (Even the removal of an end coin results in a new game for a line of shorter length.) It seems appropriate then to use an induction argument to analyze large games, noting that after a single move, the analysis is reduced to that of smaller games.

Let's make the assumption that any game involving k or fewer coins can be won if an odd number of heads is present initially. (The case $k = 1$ as we have seen is valid.) Consider a game then with $k + 1$ coins (and an odd number of heads). The first move is key. The number of heads is odd, so at least one head is in this line and a first move is possible. Choose from this line the leftmost head. It is easy to see that removing this coin and flipping its neighbors creates two (or perhaps just one) shorter line segments, each containing an odd number of heads. By the induction hypothesis these smaller games can all be won (choose the leftmost heads again), so the entire puzzle is solvable.

This completes one part of the proof. We now need to show that if the line contains an even number of heads-up coins, then the puzzle definitely cannot be won.

If the number of heads-up coins is zero, you cannot win. You cannot even make a move! Suppose at least two coins are heads-up. The selection and removal of any of these coins breaks the line of pennies into two segments. (One of these segments could be empty if we chose an end coin.) One segment must contain an odd number of heads, the other an even number. Including the selected coin itself, this adds up to an even number of heads. Upon flipping the neighboring coins we thus are assured of creating a smaller line of coins containing an even number of heads. An induction argument (like the one above) would then show that we are doomed to failure no matter which coin we initially chose!

Acknowledgments and Further Reading

Martin Gardner writes about such parity problems in [Gard2], [Gard7], and [Gard14]. A more rigorous account of Gauss's observation appears in [Rade]. For an interesting variant of this phenomenon see [Konh], problem 125. Ravi Vakil also plays a Catch Me If You Can game in his book [Vaki]. The penny solitaire puzzle is based on a circular version of the game presented in [Beck].

30 Chessboard Maneuvers

30.1 Grid Walking

Using the coloring scheme of the hint, if Lashana begins on a white cell, she will remain solely on white cells as she moves about the array. As the white cells themselves form a square grid oriented 45° to the original square lattice, we see that Lashana is in precisely the same predicament as Greta and thus will be required to take an even number of steps to complete a loop.

The following coloring scheme shows that any loop Lopsided Charlie makes must be composed of a total number of steps divisible by three!

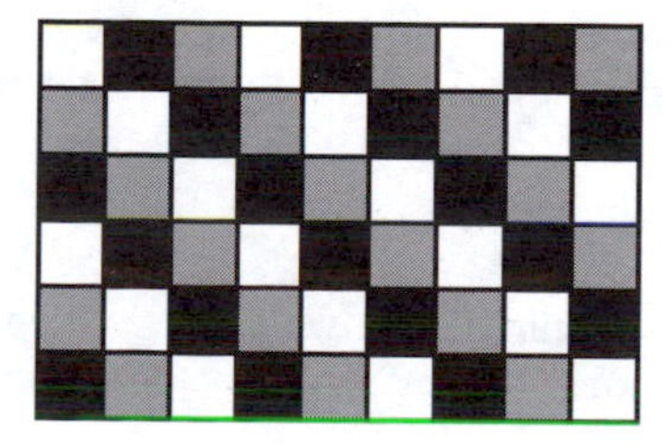

Challenge 1. Suppose instead Lopsided Charlie walked about the grid as shown.

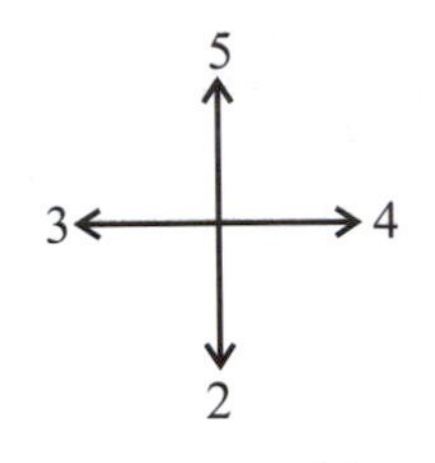

Devise a color scheme with seven colors to demonstrate that any loop Charlie now makes must be composed of a total number of steps divisible by seven.

Challenge 2. Imagine moving from room to room in a large building of cubical offices (a *cubical lattice* in 3-space). Suppose each "step" takes you to the office above or below you, in front of or behind you, or to your left or right. Can anything be said about the total number of steps you must take in a journey that eventually returns you to your original starting office? What can be said for lopsided motions?

Challenge 3. Can anything be said about loop walking on a triangular lattice?

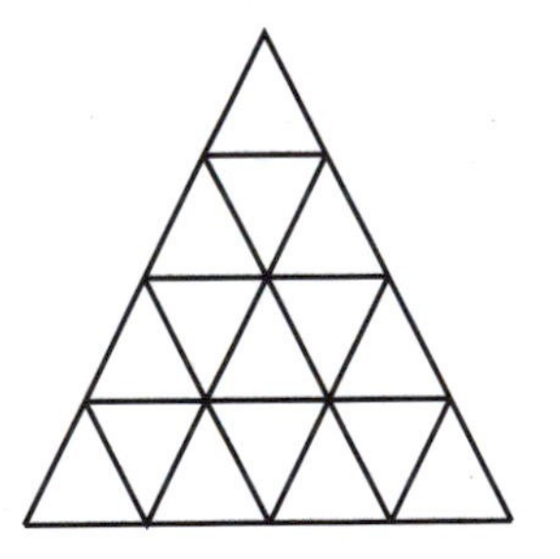

30.2 Kingly Maneuvers

Consider the analogy presented in the hint. Clearly if black fails to form a dam across the river, there must exist a stream of white water passing all the way from left to right. The nonexistence of a black path thus ensures the existence of a white path. We also deduce that the nonexistence of a white path ensures the existence of a black one by reversing the roles of black and white in this analogy. Thus one cannot simultaneously avoid paths of these two types, so every play of the two-person game must guarantee a winner.

This beautifully swift argument makes the result now seem surprisingly obvious. But are you satisfied? As a mathematician, I confess, I am not. I am aware that many "utterly obvious" statements can turn out to be false when given further thought (see, for example, section 14.2). On what mathematical principles is the above argument based? What are we assuming as standard behavior of paths in a plane?

The result is indeed true, but a more rigorous mathematical argument is needed.

30.3 Mutual Non-Attack: Rooks

The placement of a rook on a toroidal chessboard, in terms of the elimination of available squares, is no different from the case for ordinary chessboards. Thus a maximum of eight rooks can be placed in $8! = 40320$ ways on an 8×8 toroidal chessboard.

A Note on the Chess King Theorem

Suppose the squares of an $n \times n$ chessboard are arbitrarily colored black and white. The *Chess King Theorem* asserts that, no matter the coloring scheme, it is possible for a chess king to move either from the top row to the bottom row solely on black squares, or from the leftmost column to the rightmost solely on white squares. The proof of this assertion follows.

Attach to the $n \times n$ colored board four regions, R_1, R_2, R_3, and R_4, separated by four additional edges as shown. Label the end points of these edges A_1, A_2, A_3, and A_4, and color the regions R_1 and R_3 black, R_2 and R_4 white. (See the figure on the facing page.)

Beginning at the vertex A_1, imagine attempting to walk between the squares and the four additional regions from one vertex to the next, in such a way so as to always have a black

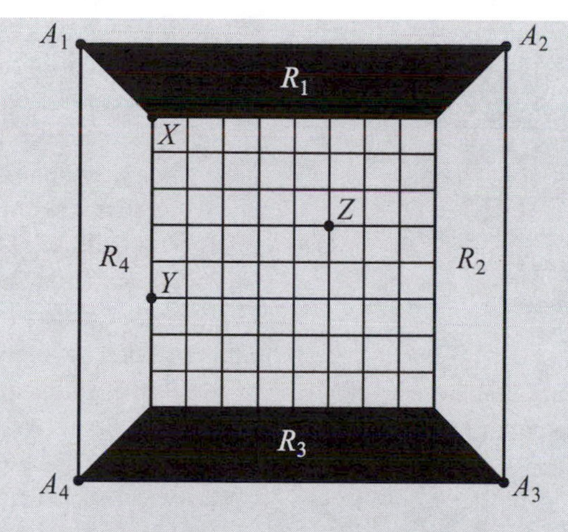

region to your left and a white region to your right. Given these requirements how far could you possibly go without ever retracing an edge?

Certainly a first move is always possible; step from A_1 to the vertex X for example. As X is at the cusp of three regions colored either black or white, there is precisely one next edge to follow. In fact, this is the case whenever we arrive at a vertex at the cusp of three regions (like X or Y): there is always precisely one edge to follow next. Also, as we are not allowed to retrace edges, we will never be able to visit these vertices more than once.

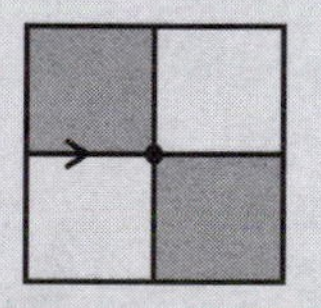

When visiting a vertex at the cusp of four regions (like Z) there may be a choice of two edges to follow next. This happens only if the colors of the regions alternate about this vertex. In this case, arbitrarily select an edge to follow. Note that it is possible to visit this vertex again at a later stage, and to follow the other edge after that, still avoiding the retracing of edges. It is clear we will never visit these types of vertices a third time.

For different coloring schemes about the vertex Z, you can quickly check that it is always possible to move along one new edge from that vertex (once we've first reached vertex Z). We will never be able to visit these vertices twice.

We have thus shown that we are always able to move one edge farther from any interior vertex different from A_1, A_2, A_3,and A_4, keeping black to the left and white to the right all the while and never retracing an edge. Because there are only finitely many edges in the diagram, such a walk must eventually stop. This can only happen if we reach A_2 or A_4. If this path lands us at A_2, then the white squares to our right represent a path for a chess king to follow from R_4 to R_2. Within this lies a path from the leftmost column to the rightmost. If we end up at A_4, we have a path of black squares to our left from R_1 to R_3 containing within it a path a chess king could follow from the top row to the bottom. This completes the proof.

Hard Challenge. Three colors are arbitrarily assigned to the cells of an $n \times n \times n$ cube. Can anything be said about the existence of paths linking opposite faces?

The placement of a rook on an 8×8 Möbius board eliminates the squares of one column and of two rows as candidates for the placement of a second rook. It follows that a maximum of four rooks can be placed on such a board. Notice that the rows of the diagram are linked together in pairs: rows 1 and 8 form one connected line, rows 2 and 7 another, 3 and 6 a third, and 4 and 5 a fourth pair. A rook can be placed in any one of the 16 cells of the first

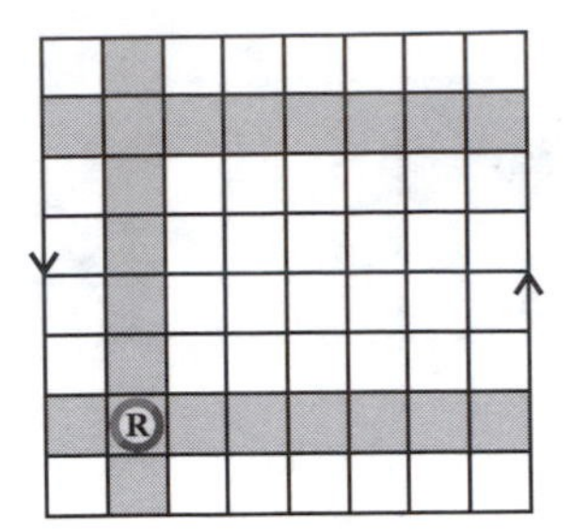

pair of rows. Once in place, there are 14 positions (avoiding the same column) in which to place the second rook; and so on. Thus there are a total of $16 \times 14 \times 12 \times 10 = 26880$ ways to arrange four rooks on an 8×8 Möbius chessboard in mutual non-attack.

Ignoring for the moment the identification of the top and bottom edges in an 8×8 projective plane chessboard, note that the rows of this diagram are linked in pairs as before. A rook can be placed in any one of 16 positions of the row 1 and row 8 pair. Given the top-to-bottom edge identification, this eliminates four cells from the row 2 and row 7 pair, and just 12 squares remain as candidates for the placement of a second rook in this second row pair. Continuing in this way we see that a maximum of four rooks can be placed on an 8×8 projective plane chessboard in $16 \times 12 \times 8 \times 4 = 6144$ different ways.

The rooks problem for a three-dimensional cubical lattice is considerably more difficult to analyze.

A Note on Rooks in a Three-Dimensional Chessboard

The placement of rooks in a three-dimensional chessboard is still an area of active research. What is the maximal number of rooks that can be placed within an $n \times n \times n$ board in mutual non-attack? The question is easy to answer, but determining the total number of different arrangements is still a challenge.

First note that each layer of a cubical chessboard operates as its own two-dimensional $n \times n$ board. We know that each layer can contain a maximum of n rooks in mutual non-attack. As there are n layers, at most n^2 rooks can be appropriately placed within the three-dimensional lattice. It is easy to devise schemes to demonstrate that the placement of n^2 rooks is indeed achievable. For example, the following array represents a configuration of nine rooks in a $3 \times 3 \times 3$ chessboard. A 1 represents a rook in the first layer; each 2 is a rook in the second layer and 3 is for the third. The diagram represents the view we would see if we looked through the cube from above.

1	2	3
2	3	1
3	1	2

In representing solutions this way, we see that our three-dimensional rooks problem is really no different from that of constructing $n \times n$ *Latin squares*, that is, square matrices with entries numbered from 1 to n and arranged so that no row or column contains two identical numbers. There are twelve 3×3 Latin squares, meaning there are twelve ways to arrange nine rooks in a $3 \times 3 \times 3$ chessboard. The following table shows how many Latin squares of order n exist for n from 1 to 10. Each row corresponds to the number of ways possible to place n^2 rooks in mutual non-attack on an $n \times n \times n$ board.

n	Number of $n \times n$ Latin squares
1	1
2	2
3	12
4	576
5	161280
6	812851200
7	61479419904000
8	108776032459082956800
9	5524751496156892842531225600
10	9982437658213039871725064756920320000

General formulas are known for the number of Latin squares of order n, but they are difficult to handle. Mathematicians hunting for more tractable analyses of the behavior of these numbers for larger values of n. See Terry Ritter's website [Ritt] for an overview of the current state of affairs.

30.4 Mutual Non-Attack: Queens

The three diagrams below show how to place the maximal number of queens in an arrangement of mutual non-attack on 4×4, 5×5, and 8×8 boards. (There are other solutions too.)

It has been established that it is indeed possible to place n queens on an $n \times n$ ($n > 3$) chessboard in mutual non-attack formation, and efficient algorithms for doing this are known. (Check out Nikolaus Shaller's website [Scha] for an overview of this topic.) If you would like to experiment more with the *n-Queens problem* check out Tim Cavanaugh's (St. Mary's College of Maryland, Class of 1999) wonderful program [Cava]. It allows you to study the problem in several fun, interactive ways.

Hard Challenge. Study the n-Queens problem on toroidal, Möbius, or three-dimensional chessboards. (Give a careful definition of "diagonal" motion in the latter case.)

Acknowledgments and Further Reading

These grid-walking puzzles also appear in my article [Tant4]. The two-person game that follows the Kingly Maneuvers puzzle is a variant of the game of *Hex* played on a rhombus-shaped board of hexagons. This game, and its amazing topological consequences, are studied in David

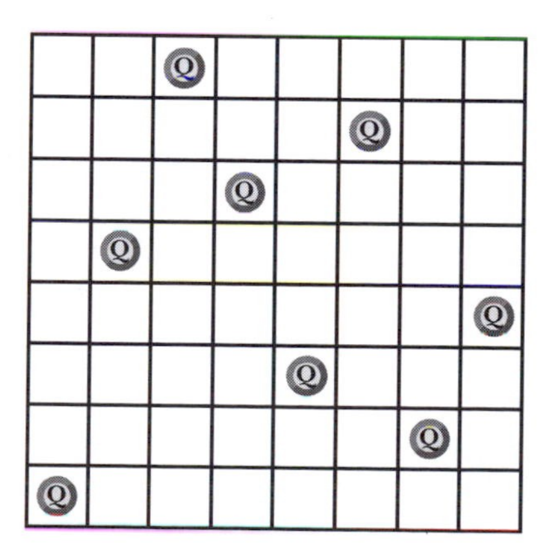

Gale's fun article [Gale1]. The proof of the Chess King Theorem I present is a modified version of his argument.

The n-Queens problem is a classic problem well known to computer scientists. It offers interesting programming challenges to students in the field. The connection between the n-Rooks problem and Latin squares is a little surprising. Latin squares can be studied in [Ball], [Dene], [Layw], and [Gard2]. An interesting application of Hall's Marriage Theorem (see section 23.2) to the theory of Latin squares appears in [Ande]. For a study of another chess problem on a toroidal board, see [Watk].

References

[Abbo] E. A. Abbot, *Flatland: A Romance of Many Dimensions*, Princeton University Press, 1991.

[Adl] I. Adler, "Make up your own card tricks," *Journal of Recreational Mathematics*, **6** (1973), pp. 87–91.

[Aign] M. Aigner, "Moving into the desert with Fibonacci," *Mathematics Magazine*, **70**, No. 1 (1997), pp. 11–21.

[Almg] F. J. Almgren, Jr., J. E. Taylor, "Geometry of soap films," *Scientific American,* **235**, No. 1 (1976), pp. 82–93.

[Ande] I. Anderson, *A First Course in Combinatorial Mathematics*, Oxford Applied Mathematics and Computer Science Series, Clarendon Press, Oxford, 1974.

[Anne] C. Anne, "Egyptian fractions and the inheritance problem," *The College Mathematics Journal,* **29**, No. 4 (1998), pp. 296–300.

[Arno] B. H. Arnold, *Intuitive Concepts in Elementary Topology*, Prentice Hall, Englewood Cliffs, NJ, 1962.

[Arti1] E. Artin, "Theory of braids," *Annals of Mathematics*, Second Series, **48**, No. 1 (1947), pp. 101–126.

[Arti2] E. Artin, "Braids and Permutations," *Annals of Mathematics,* Second Series, **48**, No. 1 (1947), pp. 643–649.

[Ball] W. W. Rouse Ball, H. S. M. Coxeter, *Mathematical Recreations and Essays,* 12th ed., University of Toronto Press, 1974.

[Barn] M. F. Barnsley, *Fractals Everywhere*, Academic Press, Inc., San Diego, CA,1988.

[Beck] D. Beckwith, Problem 10459, in Problems and Solutions, *American Mathematical Monthly*, **104**, No. 9 (1997), p. 876.

[Berl1] E. R. Berlekamp, J. H. Conway, R. K. Guy, *Winning Ways for your Mathematical Plays. Volume 1: Games in General*, Academic Press, New York, 1982.

[Berl2] E. R. Berlekamp, J. H. Conway, R. K. Guy, *Winning Ways for your Mathematical Plays. Volume 2: Games in Particular*, Academic Press, New York, 1982.

[Binm] K. Binmore, *Fun and Games: A Text on Game Theory*, Heath, Lexington MA, 1992.

[Bogo] A. Bogomolny, "Cut the knot!" `www.cut-the-knot.com`

[Bond] C. Bondi, Editor, *New Applications of Mathematics*, Penguin Books, London, 1991.

[Bram1] S. J. Brams, A. D. Taylor, "An envy free cake division protocol," *American Mathematical Monthly*, **102** (1995), pp. 9–18.

[Bram2] S. J. Brams, A. D. Taylor, *Fair Division: From Cake-Cutting to Dispute Resolution*, Cambridge University Press, 1996.

[Bred] G. E. Bredon, *Topology and Geometry*, Springer-Verlag, New York, 1993.

[Brol] D. Broline, "Renumbering of the faces of dice," *Mathematics Magazine*, **52**, No. 5 (1979), pp. 312–315.

[Bunc] B. Bunch, *Mathematical Fallacies and Paradoxes*, Dover Publications, New York, 1982.

[Cava] T. Cavanaugh, "SMC math web," `www.nsm.smc.edu/MathCS/Projects/tc/index.htm`

[Chan] G. Chang, T. Sederberg, *Over and Over Again*, The Mathematical Association of America, 1997.

[Char] G. Chartrand, *Introductory Graph Theory*, Dover, New York, 1977.

[COMA] COMAP, Inc., *For All Practical Purposes: Introduction to Contemporary Mathematics*, 4th ed., W. H. Freeman, New York, 1997.

[Conw1] J. H. Conway, R. K. Guy, "Stability of polyhedra," *SIAM Review*, **11** (1969), pp. 78–82.

[Conw2] J. H. Conway, J. C. Lagarias, "Tiling with polyominoes and combinatorial group theory," *Journal of Combinatorial Theory,* Series A, **53** (1990), pp. 183–208.

[Cour] R. Courant, H. Robbins, *What is Mathematics?*, Oxford University Press, 1941.

[Crof] H. T. Croft, K. J. Falconer, R. K. Guy, *Unsolved Problems in Geometry*, Problem Books in Mathematics, Unsolved Problems in Intuitive Mathematics, Vol. 2, Springer-Verlag, New York, 1991.

[Dayk] D. E. Daykin, "The bicycle problem," *Mathematics Magazine*, **45** (1972), p. 1

[Dene] J. Denes, *Latin Squares and their Applications,* Academic Press, New York, 1974.

[Deva1] R. L. Devaney, L. Keen, Editors, *Chaos and Fractals: The Mathematics behind the Computer Graphics*}, Proceedings of Symposia in Applied Mathematics, Vol. 39, American Mathematical Society, Providence RI, 1989.

[Deva2] R. L. Devaney, *A First Course in Chaotic Dynamical Systems: Theory and Experiment*, Addison-Wesley, Reading MA, 1992.

[Dewd] A. K. Dewdney, "Two-dimensional Turing machines and tur-mites make tracks on a plane," in Computer Recreations, *Scientific American,* **261**, No. 3 (1989), pp. 180–183.

[Doyl] P. Doyle, J. L. Snell, *Random Walks and Electric Networks*, The Carus Mathematical Monographs, No. 22, The Mathematical Association of America, Washington, DC, 1984.

[Dubi] L. E. Dubins, E. H. Spanier, "How to cut a cake fairly," *American Mathematical Monthly*, **68** (1961), pp. 1–17.

[Dunh] W. Dunham, *The Mathematical Universe. An Alphabetical Journey through the Great Proofs, Problems and Personalities*, Wiley, New York, 1994.

[Eisn] L. Eisner, "Leaning tower of the physical review," Letter to the Editor, *American Journal of Physics*, **27** (1959), pp. 121–122.

[Farr] F. A. Farris, N. K. Rossing, "Woven rope friezes," *Mathematics Magazine*, **72** No. 1 (1999), pp. 32–38.

[Fern] L. Fernandez, R. Piron, "Should she switch? A game-theoretic analysis of the Monty Hall problem," *Mathematics Magazine*, **72**, No. 3 (1999), pp. 214–217.

[Fink] D. Finkelstein, J. Rubinstein, "Connection between spin statistics and kinks," *Journal of Mathematical Physics,* **9**, No. 11 (1968), pp. 1762–1779.

[Firb] P. A. Firby, C. F. Gardiner, *Surface Topology*, 2nd Ed., Ellis Horwood, New York, 1991.

[Fran1] G. K. Francis, B. Morin, "Arnold Shapiro's eversion of the sphere," *The Mathematical Intelligencer*, **2**, No. 4 (1979), pp. 200–203.

[Fran2] G. K. Francis, J. R. Weeks, "Conway's ZIP proof," *American Mathematical Monthly,* **106**, No. 5 (1999), pp. 393–399.

[Fred] B. Frederick, J. Hersberger, "The mathematical judge: A fable," *The College Mathematics Journal*, **26**, No. 5 (1995), pp. 377–381.

[Gale1] D. Gale, "The game of hex and the Brouwer fixed point theorem," *American Mathematical Monthly*, **86** (1979), pp. 818–827.

[Gale2] D. Gale, "The industrious ant," in Mathematical Entertainments, *Mathematical Intelligencer*, **15**, No. 2 (1993), pp. 54–55.

[Gale3] D. Gale, J. Propp, S. Sutherland, S. Troubetzkoy, "Further travels with my ant," in Mathematical Entertainments, *Mathematical Intelligencer*, **17**, No. 3 (1995), pp. 48–56.

[Gard1] M. Gardner, *Mathematics, Magic and Mystery*, Dover, New York, 1956.

[Gard2] M. Gardner, *Martin Gardner's New Mathematical Diversions from Scientific American*, Simon & Schuster, New York, 1966.

[Gard3] M. Gardner, *The Scientific American Book of Mathematical Puzzles and Diversions*, Simon & Schuster, New York, 1959.

[Gard4] M. Gardner, *The 2nd Scientific American Book of Mathematical Puzzles and Diversions*, Simon & Schuster, New York, 1961.

[Gard5] M. Gardner, "Curves of constant width, one which makes it possible to drill square holes," in Mathematical Games, *Scientific American*, **208** (1963), pp. 148–156.

[Gard6] M. Gardner, *The Unexpected Hanging and other Mathematical Diversions*, University of Chicago Press, 1969.

[Gard7] M. Gardner, *The 6th Book of Mathematical Diversions from Scientific American*, University of Chicago Press, 1971.

[Gard 8] M. Gardner, "Amazing mathematical card tricks that do not require prestidigitation," in Mathematical Games, *Scientific American*, **227,** No. 1 (1972), pp. 102–105.

[Gard9] M. Gardner, *Aha! Insight*, W. H. Freeman, New York, 1978.

[Gard10] M. Gardner, *Mathematical Magic Show*, Mathematical Association of America, Washington, DC, 1990.

[Gard11] M. Gardner, *Mathematical Circus*, Mathematical Association of America, Washington, DC, 1992.

[Gard12] M. Gardner, *Aha! Gotcha: Paradoxes to Puzzle and Delight*, W. H. Freeman, New York, 1982.

[Gard13] M. Gardner, *Wheels, Life and other Mathematical Amusements*, W. H. Freeman, New York, 1983.

[Gard14] M. Gardner, *The Magic Numbers of Dr. Matrix*, Prometheus Books, New York, 1985.

[Gard15] M. Gardner, *Knotted Doughnuts and other Mathematical Entertainments*, W. H. Freeman, New York, 1986.

[Gard16] M. Gardner, *Time Travel and other Mathematical Bewilderments*, W. H. Freeman, New York, 1988.

[Gard 17] M. Gardner, *Mathematical Carnival*, The Mathematical Association of America, Washington DC, 1989.

[Gard18] M. Gardner, *Penrose Tiles to Trapdoor Ciphers ... and the Return of Dr. Matrix*, W. H. Freeman, New York, 1989.

[Gard19] M. Gardner, "Tiling the bent tromino with n congruent shapes," *Journal of Recreational Mathematics*, **22** (3) (1990), pp. 185–191.

[Gard20] M. Gardner, *Fractal Music, Hypercards and More... Mathematical Recreations from Scientific American Magazine*,W. H. Freeman, New York, 1992.

[Gard21] M. Gardner, "Ten amazing mathematical tricks," *Math Horizons,* September 1998, pp. 13–15.

[Gard22] M. Gardner, "Ridiculous questions," *Math Horizons,* November 1996, pp. 24–25.

[Gior] F. R. Giordano, M. D. Weir, *A First Course in Mathematical Modeling*, Brooks/Cole, Montery CA, 1985.

[Golo] S. W. Golomb, *Polyominoes: Puzzles, Patterns, Problems and Packings*," 2nd ed., Princeton University Press, 1994.

[Grin] C. M. Grinstead, J. L. Snell, *Introduction to Probability,* 2nd Ed., American Mathematical Society, Providence RI, 1997.

[Hake] W. Haken, "An attempt to understand the four color problem," *Journal of Graph Theory*, **1** (1977), pp. 193–206.

[Hall] N. Hall, Editor, *Exploring Chaos: A Guide to the New Science of Disorder,* W. W. Norton, New York, 1991.

[Hara] F. Harary, "The four colour conjecture and other graphical diseases," in P*roof Techniques in Graph Theory, Proceedings,* F. Harary editor, Academic Press, New York, 1969, pp. 1–9.

[Hass1] J. Hass, M. Hutchings, R. Schafly, "The double bubble conjecture," *Electronic Research Announcements of the American Mathematical Society*, **1** (1995), pp. 98–102.

[Hass2] J. Hass, "General double bubble conjecture in $\mathbb{R}^3$ solved," *Focus,* **20**, No. 5 (2000), pp. 4–5.

[Hepp] A. Heppes, "A double tipping tetrahedron," *SIAM Review*, **9** (1967), pp. 599–600.

[Hild] S. Hildebrandt, A. Tromba, *Mathematics and Optimal Form*, Scientific American Books, New York, 1985.

[Hoff] P. Hoffman, *Archimede's Revenge: The Joys and Perils of Mathematics*, W. W. Norton, New York, 1988.

[Hons1] R. Honsberger, *Ingenuity in Mathematics,* New Mathematical Library, No. 23, The Mathematical Association of America, Washington DC, 1970.

[Hons2] R. Honsberger, *Mathematical Gems I,* Dolciani Mathematical Expositions, No. 1, The Mathematical Association of America, Washington DC, 1973.

[Hons3] R. Honsberger, Editor, *Mathematical Plums*, Dolciani Mathematical Expositions, No. 4, The Mathematical Association of America, Washington DC, 1979.

[Jaco] K. Jacobs, *Invitation to Mathematics*, Princeton University Press, 1992.

[Jayn] E. T. Jaynes, "The well-posed problem," in *Papers in probability, statistics and statistical physics*, R. D. Rosencrantz, ed., (Dordrecht: D. Reidel, 1983), pp. 133–148.

[John] P. B. Johnson, "Leaning tower of life," *American Journal of Physics,* **23** (1955), p. 240.

[Kac] M. Kac, S. Ulam, *Mathematics and Logic*, Dover Publications Inc., New York, 1992.

[Kilg] D. M. Kilgour, S. J. Brams, "The truel," *Mathematics Magazine*, **70**, No. 5 (1997), pp. 315–326.

[Knut] D. E. Knuth, "The triel: A new solution," *Journal of Recreational Mathematics*, **6**, No. 1 (1973), pp. 1–7.

[Kola] G. Kolata, "Perfect shuffles and their relation to math," *Science*, **216** (1982), pp. 505–506.

[Konh] J. Konhauser, D. Velleman, S. Wagon, *Which Way did the Bicycle Go? and other Intriguing Mathematical Mysteries*, Dolciani Mathematical Expositions No. 18, The Mathematical Association of America, Washington DC, 1996.

[Lan] C. G. Langton, "Studying artificial life with cellular automata," *Physical D*, **22** (1986), pp. 120–149.

[Lang] R. Lang, "Robert Lang's report on his visit to Japan," *The Newsletter for the Friends of the Origami Center of America*, Issue No. 43 (1993), pp. 1, 8, 9, 24.

[Layw] C. F. Laywine, G. L. Mullen, *Discrete Mathematics using Latin Squares*, Wiley-Interscience Series in Discrete Mathematics and Optimization, Wiley, 1998.

[Leep] D. Leep, G. Myerson, "Marriage, magic and solitaire," *American Mathematical Monthly*, **106**, No. 5 (1999), pp. 419–429.

[Lend] J. Lendering, "Articles on ancient history," `http://home.wxs.nl/~lende045/Josephus/Josephus.html`.

[Levy] S. Levy, D. Maxwell, T. Munzer, Directors, *Outside In*, Geometry Center, Univeristy of Minnesota, 1994.

[Mass] W. S. Massey, *A Basic Course in Algebraic Topology*, Springer-Verlag, New York, 1991.

[May] K. O. May, "The origin of the four-color conjecture," *Isis*, **56** (1965), pp. 346–348.

[McGi] L. McGilvery, "Speaking of paradoxes ... or are we?" *Journal of Recreational Mathematics*, **19** (1) (1987), pp. 15–19.

[Morg] F. Morgan, E. R. Melnick, R. Nicholson, "The soap-bubble-geometry contest," *The Mathematics Teacher*, **90**, No. 9 (1997), pp. 746–749.

[Morr] B. Morris, *Magic Tricks, Card Shuffling and Dynamical Computer Memories: The Mathematics of the Perfect Shuffle*, Mathematical Association of America, Washington DC, 1998.

[Mnat] M. Mnatsakanian, "Annular rings of equal areas," *Math Horizons,* November 1997, pp. 5–8.

[Newm] M. H. A. Newman, "On a string problem of Dirac," *The Journal of the London Mathematics Society*, **17**, Part 3, No. 67 (1942), pp. 328–333.

[Orli] P. Orlik, H. Terao, *Arrangements of Hyperplanes*, Grundlehren der Mathematischen Wissenschaften 300, Springer-Verlag, Berlin, 1992.

[Papp] T. Pappas, *More Joy of Mathematics*, Wide World Publishing/Tetra, San Carlos CA, 1991.

[Paul] J. A. Paulos, *Innumeracy: Mathematical Illiteracy and its Consequences*, Hill & Wang, New York, 1988.

[Pete1] I. Peterson, "Mathematical shuffling," *Science News*, **125** (1984), p. 202.

[Pete2] I. Peterson, *The Mathematical Tourist: Snapshots of Modern Mathematics*, W. H. Freeman, New York, 1988.

[Pete3] I. Peterson, *Islands of Truth. A Mathematical Mystery Cruise*, W. H. Freeman, New York, 1990.

[Pete4] I. Peterson, "Toil and trouble over double bubbles," *Science News*, **148** (1995), p. 101.

[Pete5] I. Peterson, "A penny surprise," in *Ivar Peterson's Math Trek*, `http://206.4.57.253/mathland/mathtrek_12_15.html`.

[Phil] A. Phillips, "Turning a surface inside out," *Scientific American,* **214** (1966), pp. 112–120.

[Prop1] J. Propp, "Further ant-ics," *Mathematical Intelligencer*, **16**, No. 1 (1994), pp. 37–42.

[Prop2] J. Propp, "A pedestrian approach to a method of Conway, or a tale of two cities," *Mathematics Magazine*, **70**, No. 5 (1997), pp. 327–340.

[Rade] H. Rademacher, O. Toeplitz, *The Enjoyment of Math*, Princeton University Press, 1957.

[Ritt] T. Ritter, "Research comments from *Ciphers by Ritter*," `www.io.com/~ritter/RES/LATSQ.HTM`.

[Robe] N. Robertson, D. P. Sanders, P. D. Seymour, R. Thomas, "A new proof of the four color theorem," *Electronic Research Announcements of the American Mathematical Society*, **2** (1996), pp. 17–25.

[Röni] W. von Rönik, "Doughnut slicing," *The College Mathematics Journal*, **28**, No. 5 (1997), pp. 381–383.

[Ruck] R. Rucker, *The Fourth Dimension. Toward a Geometry of Higher Reality*, Houghton Mifflin, Boston, 1984.

[Saar] D. G. Saari, F. Valognes, "Geometry, voting and paradoxes," *Mathematics Magazine*, **71**, No. 4 (1998), pp. 243–259.

[Saat] T. L. Saaty, P. C. Kainen, *The Four-Color Problem: Assaults and Conquests*, Dover , New York, 1986.

[Scha] N. Schaller, "The N-queens problem (NQP)," `http://ourworld.compuserve.com/homepages/Nikolaus_Schaller/NQP.html`

[Shep] J. A. H. Shepperd, "Braids which can be plaited with their threads tied together at each end," *Royal Society of London Proceedings,* A, **265** (1962), pp. 229–244.

[Simm] G. F. Simmons, *Calculus with Analytic Geometry*, 2nd Ed., McGraw-Hill, New York, 1996.

[Sing] D. Singmaster, "Covering deleted chessboards with dominoes," *Mathematics Magazine*, **48**, No. 2 (1975), pp. 59–66.

[Stew] I. Stewart, "The ultimate in anty-particles," in Mathematical Games, *Scientific American*, **271**, No. 1 (1994), pp. 104–107.

[Stra] P. Straffin, *Game Theory and Strategy*, The Mathematical Association of America, Washington, DC, 1993.

[Su] F. E. Su, "Rental harmony: Sperner's lemma in fair division," *American Mathematical Monthly*, **106**, No. 10 (1999), pp. 930–942.

[Tant1] J. S. Tanton, "A dozen questions about a donut," *Math Horizons*, November 1998, pp. 26–31.

[Tant2] J. S. Tanton, "A dozen reasons why 1=2," *Math Horizons*, February 1999, pp. 21–25.

[Tant3] J. S. Tanton, "A half-dozen mathematical activities to try with friends," *Math Horizons*, September 1999, pp. 26–31.

[Tant4] J. S. Tanton, "A dozen questions about squares and cubes," *Math Horizons*, September 2000, pp. 26–30, 34.

[Tant5] J. S. Tanton, R. Robb, "On layered tilings," In preparation.

[Tayl] H. Taylor, "Bicycle tubes inside out," in *The Mathematical Gardner*, D. A. Klarner, Editor, Wadsworth International, Belmont CA, 1981.

[Thur] W. P. Thurston, "Conway's tiling groups," *American Mathematical Monthly*, **97**, No. 8 (1990), pp. 757–773.

[Toti] V. Totik, "A tale of two integrals," *American Mathematical Monthly*, **106**, No. 3 (1999), pp. 227–240.

[Vaki] R. Vakil, *A Mathematical Mosaic: Patterns and Problem Solving*, Brendon Kelly Publishing Inc., Burlington, Ontario, 1996.

[Vick] J. Vick, *Homology Theory. An Introduction to Algebraic Topology*, Springer-Verlag, New York, 1994.

[Wago] S. Wagon, *The Banach-Tarski Paradox*, Cambridge University Press, 1985.

[Watk] J. Watkins, R. L. Hoenigman, "Knight's tour on a torus," *Mathematics Magazine*, **70**, No. 3 (1997), pp. 175–184.

[Week] J. R. Weeks, *The Shape of Space*, Marcel-Decker, New York, 1985.

INDEX

Parity

Physics

Probability Theory

Shapes, Surfaces and Space

Topology

Voting, Sharing, Distributing Goods